Nanomaterials in Concrete

NANOMATERIALS IN CONCRETE

Advances in Protection, Repair, and Upgrade

Henry E. Cardenas, Ph.D.

*Associate Professor of Mechanical
and Nanosystems Engineering
Louisiana Tech University*

DES*tech* Publications, Inc.

Nanomaterials in Concrete

DEStech Publications, Inc.
439 North Duke Street
Lancaster, Pennsylvania 17602 U.S.A.

Printed in the United States of America
10   9   8   7   6   5   4   3   2   1

Main entry under title:
Nanomaterials in Concrete: Advances in Protection, Repair, and Upgrade

A DEStech Publications book
Bibliography: p.
Includes index p. 171

ISBN: 978-1-60595-050-1

To
Tracy, Matthew, Bethany, and Robbie

Table of Contents

Acknowledgements

I wish to express my gratitude to Dr. Leslie Struble, Professor of Civil Engineering, University of Illinois Urbana-Champaign (UIUC), for her advice and support, and for graciously making her time, laboratory, and information resources extensively available at the beginning of this odyssey. The generous insight of Dr. Thomas Mason, Professor of Material Science, Northwestern University in Evanston, Illinois, is especially appreciated.

Deep gratitude is expressed to the author's late father, Victor M. Cardenas, Chemical Engineer, Vanex Color of PPG Industries, Mt. Vernon, Illinois. His inspiration and extensive technical assistance in the laboratory was phenomenal. I wish he could be here to see this project completed.

The author also wishes to extend his sincere appreciation to Mr. David Hodge, President of Litania Sports, in Urbana, Illinois for the initial financial and moral support of this work. Other generous sponsors include Ron Smith, HRS LLC, Paul Femmer, Osmotec, LLC, Dr. Paul Todd, TechShot Inc., Charles Turk, Entergy Nuclear, Inc., U.S. Army Corps of Engineers, NASA Kennedy Space Center, and the A. J. Weller Corporation. Significant material and technical support that was much appreciated, was also provided by John Grim and John Hughes, Applications Engineers with Nalco Chemical. Thanks are also extended to the technical sales staff at Oxychem for provision of other test materials. Mr. Bryant Mather, past president of the American Concrete Institute and the American Society of Materials and Testing is gratefully acknowledged for his generous insight and technical information, as is also Dr. Francis Young, Professor Emeritus of Civil Engineering at UIUC, for his sage experimental advice.

The efforts of several of my past students have also contributed greatly to this project. I wish to thank Dr. Kunal Kupwade-Patil for his relentless effort, support, and enormous productivity. My sincere appreciation also goes to Aliya Arenova, Anupam Joshi, Dr. James Philips, Josh Alexander, Anjaneyulu Kurukunda, Mark Castay, James Eastwood, Syed Faisal, Nagaraju Goli, Padmanabhan Venkatachalam, Shailesh Madisetti, Oner Moral, Johnathan Needham, Israel Popoola, Lakshim Rachapudi, Shailesh Madisetti, Pradeep Paturi, Ralph Serrano, Nicholas Richardson, and Joey Zhao.

My multi-disciplinary collaborators from Louisiana Tech University have had a tremendous impact on making many of the findings in this work possible. Dr. Jinko Kanno, Mathematics, Dr. Daniela Mainardi, Chemistry, Dr. Sven Eklund, Chemistry, Dr. Sidney Sit, Biomedical Engineering, Dr. Yuri Lvov, Physics, and Dr. Luke Lee, Civil Engineering have all blessed this work with valuable contributions. From Kennedy Space Center, the Corrosion Technology Group, Dr. Luz Marina Calle, Dr. Mark Kolody, Dr. Paul Hintze, Dr. Wendy Li, Mr. Joe Curren are also deeply appreciated.

Above all, I wish to thank my wife, Tracy Cardenas, for her gracious support, understanding, and sacrifice; and my children, Matthew, Bethany, and Robbie, for their patience and inspiration.

Introduction

IT is widely noted that advances in nanotechnology will have a disruptive impact on society. Nanosystems engineering is expected to change the way we live as much as the train, the automobile, or the computer. It is remarkable that during our 5000 year history with concrete we have gone about our lives giving little thought to the surfaces and structures upon which we work, live and drive. In contrast, today's era of rapid global economic development is straining our capacity to build and sustain the infrastructure of civilization, because modern concrete manufacturing leaves an ecological footprint second only to the combustion engine. At the same time, the need to construct concrete that is strong and durable has never been greater—an imperative surpassed only by the need to sustain the durability of existing structures instead of rebuilding them. This text describes recent advances in nanomaterials processing which offer historic opportunities for minimizing the financial and ecological burdens required to maintain these essential assets.

Until now, concrete structural enhancement could only be engaged at the point of construction, while the mix is being formed. For concrete rehabilitation, repairs could only be achieved by adding on to the structure, and this often requiring partial demolition. Recent developments in nanotechnology are starting to change the rules. The fertile combination of electrokinetic processing and nanoparticle engineering has opened the door to a broad spectrum of alternative repair strategies ranging from decontamination, permeability reduction, and strength enhancement, to remediation from various types of chemical degradation processes. Remarkable structural repairs and upgrades are now possible using nanomaterials and electromutagenic processing methods that can radically reconfigure the microstructure of common building materials and provide significant service life extensions without changing the physical dimensions or the external appearance of the structure.

The book is written to address the interests of the researcher and the engineer. Chapter 1 introduces the ways in which nanomaterials are applied. Much of this section focuses on specific phase development and modification methods as influenced by particle selection and electric field application. Chapters 2 through 4 focus on the basic applications of nanomaterial modifications to concrete. These include permeability reduction, porosity reduction, strength enhancement, and crack repair. More specialized durability applications are explored in Chapters 5 through 7. These include mitigation of reinforcement corrosion, recovery from sulfate attack, and recovery from freeze-thaw damage. Chapter 8 presents the practical engineering aspects of planning, setting up, and executing a given treatment application. In Chapter 9 the concept of electromutagenic processing is introduced, and specific opportunities are described for radical changes to both concrete microstructures and the properties that depend upon them. This book represents a small step in the early progress of a growing technology. Any comments and suggestions will be greatly appreciated.

CHAPTER 1

Nanomaterial Application Methods

IN order to harness the capabilities of nanomaterials it is important to understand how they interact with the environment. Manipulating these patterns of interaction enables us to control the properties that we seek to develop. This chapter focuses on the descriptions of several basic modes of interaction between nanoparticles and the environment we seek to alter. While these topics certainly involve short range transport, a more formal development of transport concepts for nanomaterials will be covered in Section 8.1.

1.1. DOPING

Nanomaterials can be built by placing a base material in contact with other materials that are naturally attracted to the base material. For instance, a surface can be immersed in a fluid in which nanoparticles are suspended. The surface may naturally carry a negative charge. If the nanoparticles carry a positive charge, they will be attracted to the surface and stick to it by electrostatic attraction. These positive particles are the dopant in this system. Placing them in contact with the surface to which they will stick is the process of doping. This concept is illustrated in Figure 1.1, where the silica surface carries a negative charge which attracts the positively charged alumina nanoparticles.

The amount of nanoparticles that stick depends upon the size of the nanoparticles and the number of binding sites that occur on the surface. This binding can change the surface energy as well as its charge. The surface in Figure 1.2 can thus be rendered non-attractive toward additional positive particles because it already carries a positive charge.

1

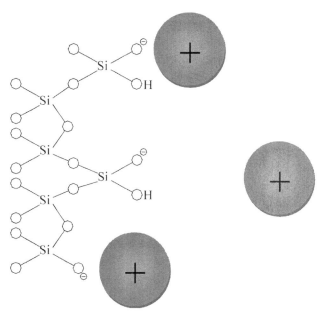

FIGURE 1.1. *Positively charged alumina nanoparticles electrostatically attracted to a silicate nanostructure immersed in water. The middle-right nanoparticle is repelled from the available bonding site due to electrostatic repulsion from the other two positive particles.*

The surface charge of a nanoparticle can also be altered in this manner. In Figure 1.2, a 20 nm particle of silica carrying a negative charge has attracted a number of 2 nm colloidal alumina particles, which have changed the surface charge of the composite particle.

Controlling the surface charge of a nanoparticle is extremely important for controlling transport. The repulsion obtained from a strong surface charge helps minimize the opportunities for particles to collide and stick together. This is referred to as flocking or coagulation. Flocks of particles are less stable in a suspension because gravity has more influence on their motion as they become increasingly massive. At some point, the amount of mass causes gravity to dominate the transport and the particle falls out of suspension.

The sign of the surface charge, positive or negative, will govern the direction of drift. This direction can have a tremendous influence on the outcome of a nanoparticle treatment. For example, if a treatment was designed to drive nanoparticles onto a structure of iron embedded in concrete or soil, a positive surface charge on the particle would mean that the electric field required for treatment would not be expected to hurt the structure. If the particle were negatively charged, then the direction of the electric field would be more difficult to handle. For the example of iron in concrete, the electric field established to drive the particles would cause the iron to dissolve fairly quickly.

The magnitude of the surface charge also influences the velocity of drift that can be obtained when the suspension is subjected to the electric field. The drift velocity that can be achieved per unit of electric field is called electrokinetic mobility. The electric field is simply the voltage potential divided by the distance between the electrodes that are in direct contact with the fluid medium.

1.2. ELECTRODEPOSITION COAGULATION ASSEMBLY

When nanoparticles are undergoing electrophoresis they can be driven toward a conductive surface and deposited onto it when they arrive. Since they are repelling each other, the particles tend to space themselves on that surface, occupying places that are not already taken. Thus electrodeposition produces a fairly uniform surface. Figure 1.3 shows large 20 nm particles of silica in the process of losing their covering of 2 nm alumina particles. These smaller particles carry a positive charge and are now able to separate from the silica carrier particle and proceed to the metal surface. The combination of electrostatic attraction to the surface and repulsion from the neighboring particles yields a uniform pattern of electrodeposited spheres. Attraction usually wins the competition, forcing the particles close enough together so that they begin coagulating as a result of van der Waals forces. In general, these forces are

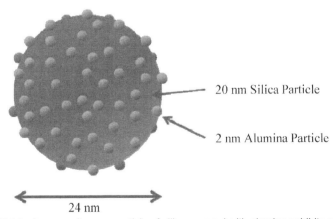

20 nm Silica Particle

2 nm Alumina Particle

24 nm

FIGURE 1.2. A composite nanoparticle of silica covered with alumina exhibits a positive surface charge. Illustration developed by Dr. Sven Eklund, Louisiana Tech University.

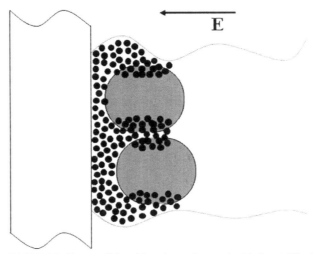

FIGURE 1.3. Nanoparticles driven toward an embedded metal flock at the surface, where they form a densely packed electrodeposited phase.

described as weak attractions that exist between molecules. They are at least strong enough to have some influence on the location of a given molecule, atom, or nanoparticle. As in this case they can also manifest between atoms that are located on the surfaces of two nanoparticles that had come into close proximity.

In the case of polymeric particles an additional assembly step is possible. If the system is permitted to dry, polymer chains can uncoil from these particles and begin to interact with neighboring particle

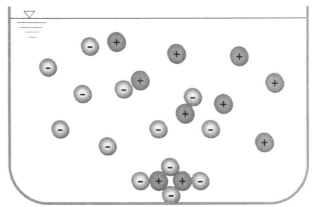

FIGURE 1.4. Oppositely charged particles are drawn together via electrostatic attraction. They form electrostatic assemblies that fall out of suspension and drop to the bottom of the beaker.

chains. Over time (hours) this particle assembly can convert itself into a uniform polymer film in much the same way as an organic paint coating.

1.3. ELECTROSTATIC ASSEMBLY

If a fluid contains nanoparticles of both positive and negative net charge, they will be attracted and readily stick to each other, just as observed earlier in the description of coagulation. The sticking force in this case tends to be stronger, since it involves both the electrostatic force of the opposing charges as well as the van der Waals forces noted earlier. In Figure 1.4 we see a beaker filled with water. Nanoparticles were added to the beaker, negative ones on the left and positive ones on the right. As the opposing charges draw the particles toward each other, some begin to stick. More particles are added to the coagulated mass until it is too heavy to remain suspended and sinks to the bottom of the beaker (precipitation).

Electrostatic assembly may also involve ions of one charge and particles of the opposite charge. When ions are involved, the opportunity for chemical reaction is significant. Nanoparticles can come under attack. In some cases they may start to dissolve, at which point their own ionic residue becomes available for the formation of new compounds that exhibit ionic or covalent bonding.

1.4. SINTERING

This assembly method is typically referred to as powder metallurgy. It is generally conducted in a dry environment in which the particles are subjected to elevated temperature. As shown in Figure 1.5, the contact

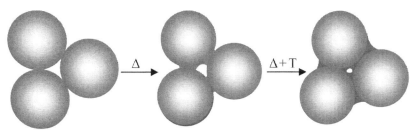

FIGURE 1.5. *Sintering process starts as temperature is elevated. The higher kinetic energy state of the atoms enables them to diffuse to the points of inter-particle contact. The surface area of the system is reduced during this process, and with it the surface energy.*

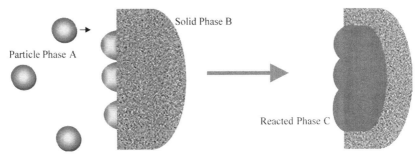

FIGURE 1.6. Nanoparticles approach a surface where they adhere and react with the substrate to form a new phase at the surface.

points between the particles are the locations where assembly occurs. Diffusing atoms migrate to these points and form chemical connections to the neighboring particles. The process is driven by the reduction in systemic surface area (and thus in surface energy) that occurs when individual particles are melded together.

1.5. REACTIVE CONVERSION

When reactive species of particles arrive at a surface, they can cause a conversion of the surface material to another phase. This process is illustrated in Figure 1.6. A given reaction can be fairly rapid, producing a new phase with distinctive properties. The depth of penetration of this new phase is not expected to be very great. In order for such conversion to occur, the particles must be attracted to the surface by an opposing charge, and then must dissolve at least partially so that ionic derivatives can react with the wall surface. In some cases, the rate of particle transport can be quite short compared with the rate of reaction.

If the reactive surface is along a pore, and the particles are small enough to penetrate it, the reaction may occur deeper within the material. It may occur all along the pore where the reactive wall phase is present. In any location where the pore becomes too narrow to permit particle passage, the penetration will be limited to that point. Other aspects of pore assembly dealing with poor formation are noted in the following section.

1.6. PORE ASSEMBLY

All the assembly processes described above can be accomplished within pores. In some cases the pores can be prepared to make room by modifying the pore system prior to the insertion of nanoparticles or other reactive species. This extension of pore volume can be viewed as a damaging process, since pore formation would take place through micro-cracking and material dissolution. In contrast, the refilling of this additional porosity with new phases that are of greater mechanical strength or chemical resistance can constitute an opportunity for enhanced material performance.

1.6.1. Electrochemical Boring

When an electric current passes through a material, the charge transport from a solid to a fluid medium can cause material to dissolve and to be carried through the fluid. In porous material this activity can take place in a pore. This is shown in Figure 1.7. Charges leaving the pore surface will do so as material within the pore is removed. This removal can be localized to phases that are relatively easy to dissolve or to locations in the material that are at a relatively high energy state due to local strain or chemistry. When this material dissolution is so localized, a pore can form or an existing pore can be extended. When it occurs in concrete, the material removed is largely calcium liberated from calcium-bearing phases.

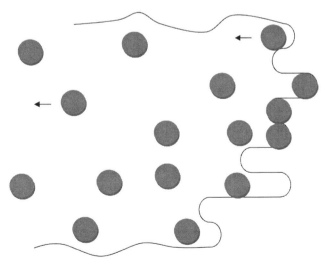

FIGURE 1.7. *Electrochemical boring extends a pore by causing material within it to jump from the solid state to the ionic state, and to be transported out of the pore.*

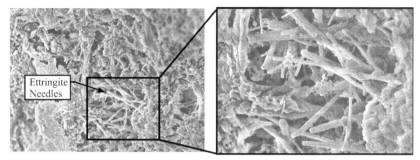

FIGURE 1.8. *The reactive growth of ettringite needles in concrete provides an increase of porosity that is usually unwelcome, but can also provide opportunities for insertion of beneficial species. Image adapted from Kupwade-Patil, K. (2010). Mitigation of chloride and sulfate based corrosion in reinforced concrete via electrokinetic nanoparticle treatment, Ph.D. Thesis, Louisiana Tech University, Ruston, LA.*

1.6.2. Reactive Pore Formation

As noted earlier, reactive ionic species can enter the pores of a material and interact with the interior surfaces. This reactivity can produce radical changes to the morphology of the micro/nanostructure. Where new structural geometry is achieved there is the opportunity to increase the porosity of the material. Figure 1.8 contains a scanning electron micrograph of concrete that has been exposed to a solution of calcium sulfate. The sulfates react with the monosulfate phases native to the concrete to produce prismatic needles of ettringite. The growth of these needles open up the pore structure—certainly a damaging occurrence, with an accompanying drop in concrete strength. Examined from another perspective, however, this change presents an opportunity to make room for the transportion of other species into the structure, resulting in a permeating alteration of properties. The pore structure, thus opened, could permit the electrokinetic injection of beneficial species. These may include polymeric particles, or even precursors of them, that could ultimately produce a material exhibiting ductility—which in turn promises better impact or vibration resistance. It could also translate into lower water-permeability and higher strength.

CHAPTER 2

Permeability Reduction

NOW that some of the fundamental modes of nanomaterial interaction have been introduced, we can explore their applications.
When a pore system is filled with particles the changes can be dramatic.
If we disregard chemical reactions for the moment, simple nanoparticle
injection provides a tremendous opportunity for rapid porosity reduction, which translates directly to strength enhancement and permeability reduction. When permeability is reduced, the benefit to durability
due to long-term savings of maintenance resources is often very great.
This chapter provides a detailed examination of the ways that we can
use nanomaterials to provide this wealth of benefits.

2.1. PERMEABILITY AND THE WET BASEMENT

In work that started as early as 1996, colloidal nanoparticles were
electrokinetically transported into hardened cement paste pores, where
they underwent chemical reactions resulting in reduced permeability.
Cardenas and others started by combining 20-nm silica and 2-nm alumina particles with simulated pore fluids to assess precipitate formation
(Cardenas and Struble, 2006, 2008). One precipitate was C–S–H, the
hydration product and binder material that is native to Portland cement
paste. The chemical formula for this phase is somewhat variable. It is
typically reported as $3 \cdot CaO \cdot 2Si_2O \cdot 8H_2O$ (Mindess *et al.* p. 58, 2003).
Permeability tests were conducted to study how these processes impact-
ed hardened cement pastes of high water/cement ratio and both high
and low alkali metal contents. Hardened cement paste in 2-inch sections

was subjected to water pressure equivalent to that which often causes residential basements to flood. It was found that 5-minute treatments using 5 volts of electrical potential difference applied over a span of 0.15 m (5 inches) was sufficient to drive nanoparticles into the pore system, even against the flow of water. The coefficient of permeability was reduced by 1 to 3 orders of magnitude, with the result that a wet basement wall, after a treatment of 5 minutes, would be dry. The following sections describe some the historical developments and the technical understanding of transport that led to this remarkable discovery.

2.1.1. Background

Water seepage through cementitious materials has been addressed using coatings, reactive grouts, and the use of pulsed electric current (Xypex, 1983; Vandex, 1983; and Mindess, Young and Darwin, 2003, p. 493). Electric current has been used to induce electroosmosis in a direction that opposes seepage into a basement (Finney, 1998). In general, electroosmosis is used concurrently with standard procedures such as chiseling out faulty joints and the application of grouts and sealers throughout the structure to close gross cracks (Hock, McInerney, and Kirstein, 1998). It is becoming clear that application of such standard techniques alone may have been sufficient to mitigate seepage in the majority of cases. It is also clear that limited understanding of the fundamental transport processes has contributed to this uncertainty.

The use of electricity to move ions into porous materials has been contemplated for some time now. The concept may first have appeared in print in England (Gratwick, 1974, p. 163), when the pumping of reactive agents into the soil side of a leaking basement was followed by the application of voltage. Similar thinking went into a patented process involving the electrokinetic transport of reactive pore-blocking agents that chemically react with some elements along the flow path (Ortlepp, 1992). Ortlepp claimed to influence the pore structure and the chemistry of paper products. Later, in Japan (Otsuki *et al.*, 1999; and Ryu and Otsuki, 2002), a similar idea applied to crack repair found some success in reducing the permeability of damaged concrete.

Other work on crack repair has examined mixtures of silica fume and calcium hydroxide for use as a mechanically applied remedy (Kasselouri, Kouloumbi, and Thomopoulos, 2001). This last approach utilized pozzolanic chemistry. Pozzolans have been mixed into concrete for decades, and their reactivity with calcium hydroxide is fairly well understood. They are inorganic microparticles that are high in SiO_2 or alumino-silica. They tend to undergo some level of hydration. They are also non-crystalline and will react with calcium hydroxide to form a hydrated phase. Most pozzolans are available as by-products of industrial processes such as the burning of coal. Examples of these are pulverized fuel ash (fly ash) and microsilica (silica fume). In 1996, the author and Dr. Leslie Struble at the University of Illinois thought about finding pozzolans small enough to fit into the pores of hardened concrete. The first positive results on permeability reduction in hardened cement paste began to be realized in the spring of 2000 (Cardenas, 2002).

In order for successful nanoparticle transport, a continuous water path needs to exist in the concrete pores. This continuous water path provides the opportunity for nanoparticle transport to occur, but it is also one of several obstacles that can hinder transport, because once a significant degree of water saturation has been established in concrete, the application of applied pressure can cause the development of an opposing laminar flow as described by Darcy's law. Fundamentally, Darcy's law is expressed as (Darcy, 1956):

$$v = Ki \tag{2.1}$$

where v is the flow velocity, K is the hydraulic coefficient of permeability, and i is the pressure gradient provided by water. Porous materials can also host other transport phenomena that are related to electrokinetic treatment, including ionic conduction, electrophoresis, and electroosmosis. Electrophoresis is the process that causes nanoparticles to move through water when an electric field is provided. Electroosmosis is an independent movement of the water inside the capillary pores. These and other phenomena related to electrokinetic transport of nanoparticles are described in detail in Section 8.1.

During early research into these questions, a significant concern was that colloidal suspensions of nanoparticles tend to become unstable when the ion content of the fluid medium gets too high (Cardenas, 2002). In concentrated solutions, the electroosmotic phenomena are suppressed because the ions in the bulk fluid cause the net charge of the particle to change. This change in the net charge can cause the particle to drop out of suspension, which would make rapid transport into the pore structure impossible to achieve without using pressure. Since pressure applications in the field are difficult at best, the hope of obtaining useful nanoparticle transport was at stake. All that was left to do was to hope that the nanoparticles would remain stable enough to continue moving after they had entered the pores of hardened cement paste.

Inside hardened cement paste, the structure of the pores dominates every aspect of nanoparticle transport (Hooton, 1988). Pores develop when water, anhydrous cement grains, and fine and coarse aggregate are mixed together. Capillary pores have the most influence on transport processes (Hearn, Hooton, and Mills, 1994). The pore characteristics that influence transport specifically include the total pore volume of the sample, distribution of pore sizes, tortuosity, and the connectivity of the pores. In addition to capillary pores, features such as microcracks and bleed paths also provide relatively permeable routes that control the overall transport rates in concretes that otherwise show low permeability (Bazant, Sener, and Kim, 1987; Brown, Shi, and Skalny, 1991; Samaha and Hover, 1992; Ludirdja, Berger, and Young, 1989; and Lim, Gowripalan, Sirivivatnanon, 2000).

In this effort it was important to identify nanoparticle species that could react with ions in the concrete pore fluid and form precipitates exhibiting low porosity. The candidates that the author considered were similar to materials that react in the presence of calcium hydroxide (Cardenas, 2002). These types of materials are used as hydraulic cements or as mineral additions for mixing concrete. Two general categories exist: pozzolans and granulated blast furnace slag. When used in the mix or later as an electrokinetic treatment, the chemical properties of these materials influence paste curing, heat evolution, early strength, durability, and permeability.

The typical pozzolanic reaction is with calcium hydroxide in the presence of water, forming calcium silicate hydrate or calcium aluminate hydrate. These reactions are much slower and less exothermic than the primary hydration reactions of Portland cement (Taylor, 1997, p. 280). In the case of fly ash the reactivity is a function of the structure of the particles. Some fly ashes have dense glassy surface layers requiring pre-treatment with basic solutions that break down surface layers and expose the interiors for useful reactivity (Fan *et al.*, 1999). In some cases, the presence of sulfate was noted as a means of accelerating pozzolanic activity (Uchikawa, 1986; Shi and Day, 2000; and Paya *et al.*, 2001).

A unique part of this project was the attempt to drive sealant nanoparticles upstream against a significant seepage of water flowing through the capillary pores. The pressure of this seepage flow was comparable to that of a commercial or residential basement. Another unique aspect of this project was the delivery of a reactive precursor to form a sealant that is chemically similar to the original binder material of the cement.

2.1.2. World's First Permeability Reduction Attempt Using Nanoparticles

This was an extremely difficult way to conduct the world's first electrokinetic nanoparticle treatment (Cardenas and Struble, 2006). An effective treatment required that the nanoparticles be transported into the pores of the hardened cement paste without piling up at the entrance and simply falling out of suspension due to the high ion concentration present in the pores. In addition, these particles needed to move fast enough to swim upstream, overcoming a flow of water that could not be very accurately estimated at the time. Assuming that these two barriers could be breached, the next challenge had to do with reactivity. It was critical that the particles should react quickly enough to keep from getting washed back out by the opposing flow. Overcoming all these problems meant that the transport characteristics of the nanoparticles and the capillary pores of hardened cement paste needed to be determined. The reactivity of the nanoparticles combined with simulated pore fluids were also examined, in order to assess the impact that these reactions would have on transport and permeability reduction within the pores.

The primary transport characteristic of the nanoparticles was electrophoretic mobility. Mobility is the velocity that the particles can attain per unit of applied voltage as distributed over (divided by) the distance between the two poles of the driving circuit. A more detailed discussion of mobility is provided in Section 8.1.2.7. Measurements were conducted to determine the electrophoretic velocities of silica and alumina nanoparticles. Velocities were obtained using Tyndall effect electrophoresis.

Electroosmosis can help or hurt the process of nanoparticle transport. Measurement of electroosmosis was conducted directly on companion specimens that were developed with the permeability flow test specimens described in the following section. These tests were run 4 weeks after each specimen was batched.

Electroosmosis was measured using the test cell shown in Figure 2.1, with some modifications. The graduated cylinders were adjusted to a total height of 0.3 m each. Copper drive electrodes were inserted along the inside of each cylinder, which permitted the easy release of gases that accumulated at each electrode. The electrodes were connected to a

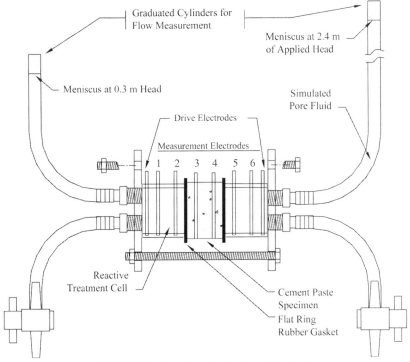

FIGURE 2.1. Reactive electrophoretic test cell.

power supply and an ammeter in series. Each specimen was saturated with high-alkali simulated pore fluid (the ingredients are listed in Table 2.1). Because high-ion content can reduce electroosmotic efficiency, it provided a worst case for observation of electroosmosis. During the test, 3 volts of DC power were applied across the drive electrodes. Prior to recording the riser fluid level, manual agitation was conducted to dissipate hydrogen bubbles that tended to form at the cathode.

All tests were conducted using a simulated pore fluid solution in-

TABLE 2.1. Composition of Low-alkali Simulated Pore Fluid for 28-day Old Paste.

Constituent/Source	Chemical Symbol	Concentration (mol/l)
Sodium hydroxide	NaOH	0.04 (Na^+)
Potassium sulfate	K_2SO_4	0.05 (SO_2^{-2})
Potassium hydroxide	KOH	0.11 (K^+)
Calcium hydroxide	$Ca(OH)_2$	< 0.002 (Ca^{+2})

TABLE 2.2. Composition of High-alkali Simulated Pore Fluid for 6-month Old Paste.

Constituent/Source	Chemical Symbol	Concentration (mol/l)
Sodium hydroxide	NaOH	0.16 (Na^+)
Potassium sulfate	K_2SO_4	0.05 (SO_2^{-2})
Potassium hydroxide	KOH	0.55 (K^+)
Calcium hydroxide	$Ca(OH)_2$	< 0.002 (Ca^{+2})

stead of just water. Simulated 6-month-old paste and pore fluid compositions were used because the chemistry of a concrete pore becomes stable after about 6 months. Tables 2.1 and 2.2 contain ingredients that represent a low-alkali, 28-day old paste, and a high-alkali, 6-month paste, as described by Taylor, 1997. These solutions were prepared using de-ionized water. The other chemicals were of reagent grade. Excess calcium hydroxide (which formed a precipitate) was included to simulate the pore fluid environment by maintaining an available source of calcium ions that dissolved out of the precipitate if the ions in the fluid became depleted due to reaction with silica- or alumina-based nanoparticles.

High permeability pastes were produced by using Portland, Type I cement manufactured by Essroc Italcementi Group in Logansport, IN, USA. The pastes were formed with high water-cement (w/c) ratios. When the w/c ratio exceeds 0.7 the percolation limit of capillary pores has been reached. This means that there is enough water to allow the pores to form a continuous pathway throughout the structure. Extra care was taken to use w/c ratios above this limit. Specimens were made using w/c ratios of 0.8 and 1.0. Ponding of fluid is a big issue when the w/c ratio gets high. For this reason, the molds needed to be sealed and rotated at 0.25 Hz for 3 days to avoid ponding. When ponding occurs it is impossible to say what the actual w/c ratio of the hard phase has become.

After demolding, the specimens were limewater-cured for 3 days. The specimens were sealed into the reactive electrophoretic test cell (shown in Figure 2.1) using 5-min epoxy. Several sets of copper electrodes (numbered 1 through 6) were installed to monitor resistivity changes. The two electrodes at the far ends of the series were used to apply driving potentials for nanoparticle transport. Each cylinder specimen was covered with 3 coats of masonry sealant. A layer of 5-min epoxy was applied 24 hours after the third coat. This reinforced the joint between the cell and the specimen at each end. With epoxy providing a sound seal, clamping forces could thus be minimized. This meant that no external loads could cause pore sizes to be reduced, altering the flow of fluid or particles. A torque of 0.3 N·m was applied to each of the four 1/4-20-threaded rods before the epoxy was applied. The specimens were allowed to sit for 8 hours to cure before the initiation of flow tests. Each permeability test was started 14 days after the specimens were cast.

The permeability flow tests were run for 3 weeks. Pressure for each flow was provided by 2.4 m of head using the high-alkali simulated pore fluid (described earlier). This prevented the solution inside the cement pores from being diluted by a flow of pure water. Water tends to absorb carbon dioxide, causing the pH of the solution to drop. To prevent this, a 5-mm layer of motor oil was placed at the meniscus of the fluid. Fluid flow was monitored using graduated pipettes that could be read to the nearest 0.01 ml at the high and low head terminals of the each setup. Corresponding fluid heights at these locations were also recorded so that the difference in head pressure could be recorded.

Five volts of DC power was applied across the drive electrodes. An appropriate polarity was applied, depending upon the zeta potential of each species being driven into the specimen face—for example, silica nanoparticles were driven into the specimen face using the negative pole. Each nanoparticle treatment was conducted for 5 minutes. Following that, each cell was rinsed 3 times with de-ionized water followed by a similar rinse with the high-alkali simulated pore fluid. At this point, the flow test was resumed to evaluate treatment impact.

The silica nanoparticles were Nalco 1050 Colloidal Silica with a 50 weight percent suspension of 20-nm silica particles. They were manufactured by Nalco Chemical Inc. of Naperville, Illinois. The alumina nanoparticles were Nalco 8676 Colloidal Alumina with a 10 weight percent suspension of 2-nm alumina particles.

2.1.3. Nanoparticle Velocity and Cement Electroosmosis

In order for the nanoparticles to move upstream a key question was their electrophoretic mobility. Mobility is the speed of travel per unit of electric field. For an electric field of ~40 V/m, the colloidal silica particles exhibited an electrophoretic velocity of 2.8×10^{-8} m/s. Colloidal alumina was faster, with a value of 1.3×10^{-7} m/s. The difference was approximately a factor of 5. It may be partially due to the large difference in size, the colloidal silica particles (20 nm) being 10 times larger than the alumina particles (2 nm). Because the drag force that opposes particle motion is proportional to this size, the alumina particles experienced a lower drag force. This was at least part of the reason why they exhibited a faster rate of electrophoresis. Nanoparticle motion was opposed by electroosmosis in the cement pores.

The findings of some electroosmosis measurements are shown in Table 2.3. In each case the electroosmotic coefficient of permeability was found to be on the order of 1 μm/A·s. How this flow influences the transport of nanoparticles depends upon the flow direction and that of the nanoparticles. For negatively charged nanoparticles this flow would be expected to provide opposition. This is because electroosmosis travels in the same direction as positive ion flow when the voltage is applied. Later in Section 8.2 more detailed analysis will explore these and other transport questions.

2.1.4. Permeability Reductions

A key motivation in this project was to investigate reactive nanoparticle electrophoresis as a means of reducing the permeability in cement. The permeabilities of untreated cement specimens were tested. At 4 weeks of age it was found that the 0.8 water-cement ratio pastes showed permeability coefficients ranging from 10^{-8} to 10^{-10} m/s. Cements with w/c = 1.0, but of similar age, had coefficients of 5×10^{-7} to 10^{-9} m/s. As a comparison, a 0.5 w/c, 1-day-old, fresh paste was at 10^{-8} m/s (Mindess, Young, and Darwin, 2003, p. 546). This same source reported a percolation limit value of 10^{-13} m/s. With this baseline behavior established, each nanoparticle treatment was conducted on the specimen face adjacent to the reactive treatment cell shown in Figure 2.1. Treatment was conducted 2 weeks after the start of each flow test, while the permeability tests were still ongoing, with 2.4 m of applied head driving the pore water in opposition to the direction of treatment.

TABLE 2.3. Electroosmosis Results.

Cement	w/c Ratio	Electroosmotic Coefficient of Permeability (μm/A·s)	Test Age (Days)
Low-alkali	0.8	1.5	31
High-alkali	0.8	5.7	29
High-alkali	1.0	1.4	45

Figure 2.2 displays the hydraulic coefficient of permeability over time. Each sample exhibited a decreasing trend, which makes sense in light of the reduction in porosity that occurs as hydration proceeds. On day 28, the control specimens that were batched with the test specimens were tested for porosity. The 0.8 w/c ratio specimens exhibited capillary porosities in the range of 25 to 28%. Similar specimens with 1.0 w/c ratios ranged from 33 to 34%. After 14 days of permeability tests (the specimens were 3 weeks old), each specimen received electrokinetic alumina nanoparticle treatment on the surface from which the water was flowing. After treatment, the specimens with 0.8 w/c ratios exhibited substantial drops in permeability. As indicated by the arrows at the lower right of the figure, the permeability was lower than could be detected with a simple gravity fed apparatus. The 1.0 w/c specimens exhibited no change in permeability apart from the expected decline over time.

The specimens in Figure 2.2 that exhibited permeability reductions had starting values of $\sim 4 \times 10^{-10}$ m/s. Those that experienced no reduction had starting values of $\sim 1 \times 10^{-8}$ m/s. The tests revealed a permeability limit at which nanoparticle velocities cannot overcome the outflow of water. It appeared that a larger driving voltage might enable penetration by nanoparticles into specimens that have higher outflows. Figure 2.2.

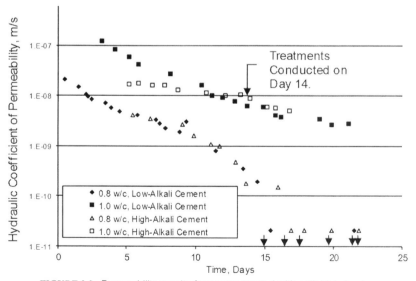

FIGURE 2.2. Permeability results for pastes treated with colloidal alumina.

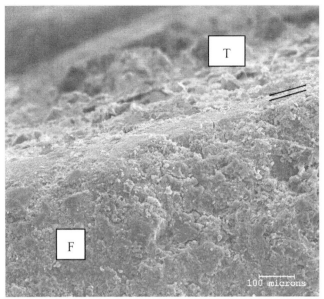

FIGURE 2.3. *SEM image of fracture surface of low-alkali cement paste (0.8 w/c) electrophoretically treated with colloidal alumina during flow test.*

2.1.5. Nanoparticle Penetrations

Specimens were fractured and examined using scanning electron microscopy (SEM) to assess any visual indication of treatment. Figure 2.3 shows a fracture surface treated with 2-nm colloidal alumina. This specimen was batched with low-alkali cement and a 0.8 w/c ratio. The top side of the specimen is the surface through which the nanoparticles entered and the water flow exited. A thin depth of ~10–20 μm on the fracture surface appeared to exhibit a more dense material with a lower apparent porosity. Parallel lines on the right side of the image demarcate the boundaries of this treated region. The untreated region shows typical porosity and shrinkage cracks that would be expected for this material. This apparent region of treatment correlates to the permeability observed in Figure 2.2. Figure 2.4 shows another SEM of the fracture surface of an untreated control specimen. This specimen exhibited no areas of lowered porosity. In Figure 2.5, an SEM image of another fracture surface of 0.8 w/c, high-alkali paste treated with 2-nm colloidal alumina, reveals that in this case the treatment only appeared to penetrate about 10–15 μm. Here again, the apparent penetration cor-

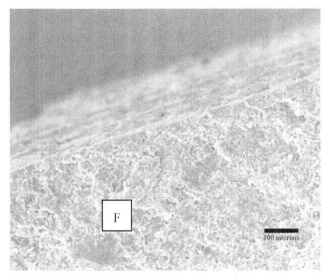

FIGURE 2.4. *SEM image of fracture surface of low-alkali cement paste control specimen (0.8 w/c).*

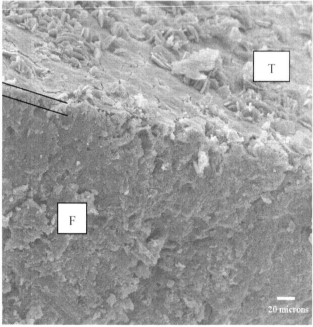

FIGURE 2.5. *SEM image of fracture surface of high-alkali cement paste (0.8 w/c) treated with colloidal alumina.*

relates to the radically reduced permeability shown for this specimen in Figure 2.2.

2.1.6. Stages of Nanoparticle Reactivity

Twelve hours after the treatments took place there was no impact on permeability. This was quite unexpected. Not until the apparatus was about to be dismantled the next day was it observed that the flows in the graduated cylinders had stopped. From the time the treatments took place in Figure 2.2, the impact was 1–2 days. In contrast, simply mixing a treatment candidate with simulated pore fluid was found to produce an immediate precipitate in less than 60 seconds. Arguably, this rapid precipitate may have been a simple coagulation of particles. One would think that this response would be insufficient to reduce permeability. It appears that this initial response keeps the collection of coagulated nanoparticles from being flushed out of the cement pores by the opposing flow of fluid. The eventual permeability reduction was likely triggered by a secondary reaction. The product of this second reaction (perhaps a pozzolanic one) is apparently yielding the dense phase needed to reduce permeability. In general, it is not surprising that a given pozzolanic reaction would be slower than some of the hydration reactions associated with Portland cement. In Portland cement, silica fume is associated with high early strength, as are some nanoscale pozzolans that have been incorporated into the mix. The difference here is that the exothermic nature of all the hydration reactions is probably providing additional thermal energy to accelerate those reactions. No such additional energy was available in these nanoparticle treatments, thus yielding the unexpectedly slow result.

2.1.7. Nanoparticle Penetration and Opposing Flow

Figure 2.6 contains a plot of the hydraulic permeability tested for specimens treated with 20-nm colloidal silica, and an additional short treatment with and calcium sulfate. Only one treatment exhibited a substantial reduction in permeability. This was the 0.8-w/c paste with low-alkali cement. The cases involving a 1.0 water/cement ratio also exhibited little or no influence from the treatment. Here again, the specimens that responded to treatment exhibited lower starting permeabilities than those that were unaffected by the treatment. The high w/c ratios appeared to provide a higher velocity of water flow opposing the transport

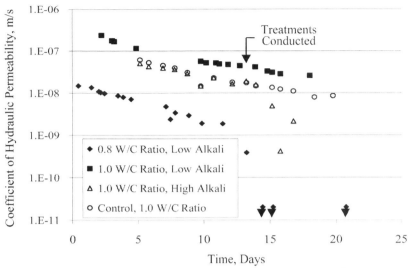

FIGURE 2.6. *Colloidal silica with calcium sulfate treatment applied to HCP during flow tests.*

of the nanoparticles. Section 8.2.2.4 describes how this penetration failure can be predicted analytically.

2.1.8. Dense Phase Formation

Figure 2.7 shows an SEM image of the single specimen from Figure 2.6 that exhibited a drop in permeability following treatment with the 20-nm colloidal silica. Here again this image exhibited a shallow region of apparent penetration adjacent to the fracture surface. The porosity in this narrow strip appears reduced as compared to deeper locations along the fracture surface. X-ray diffraction was conducted on a sample in which simulated pore fluid was mixed with the 20-nm colloidal silica particles and an excess of calcium hydroxide. The diffraction pattern shown in Figure 2.8 indicated C–S–H formation. It was a fairly close match to Plombierite, which is a variant of C–S–H. The depth of penetration in this was approximately 50–70 μm. This correlates to the large drop in permeability observed Figure 2.6.

2.1.9. Permeability Reduction in Summary

In this section of the chapter we have explored and demonstrated the viability of conducting reactive nanoparticle treatment to reduce

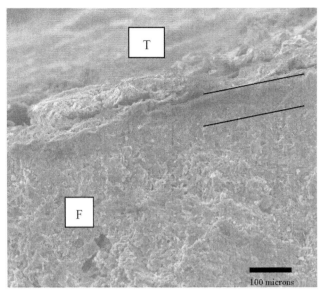

FIGURE 2.7. SEM image of fracture surface of low-alkali cement paste (0.8 w/c) electrophoretically treated with colloidal silica during flow test.

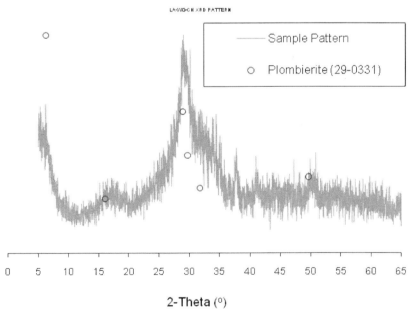

FIGURE 2.8. Powder XRD pattern of sodium silicate combined with calcium hydroxide and low-alkali simulated pore fluid showing peaks for identified phases.

permeability. Powder XRD analysis showed that 20-nm colloidal silica and pore fluid would combine with calcium hydroxide to yield a variant of C–S–H. This appears to have been one of several key reactions that actually caused permeability to drop. The principal achievement was that the reduction of permeability took place even while the cement was actively seeping fluid in direct opposition to the direction of treatment. These permeability reductions were over an order of magnitude. The delay in permeability reduction was probably due to the onset of the pozzolanic reactions that produced denser phases following the original coagulation that took place as particles entered the pores. Initial coagulation was important, because it prevented the nanoparticles from being washed back out of the pores: this allowed the necessary time for the pozzolanic reaction to yield the effective phases needed for effective permeability reduction, despite the thin depth of penetration attempted.

2.2. RADICAL PERMEABILITY REDUCTION

In this section we explore advances in permeability reduction. Treatment time was extended, each stage being 30 minutes. As before, the calcium sulfate treatment was conducted in 2 stages: first the insertion of sulfates, then the insertion of calcium ions while the sulfates were being extracted. Another point of interest explored in this section is the use of circuit resistance changes during treatment to predict the impact on permeability. The treatment candidates used in this project are listed in Table 2.4. Rhoplex is a Rohm and Haas product, described as an acrylic co-polymer. Another interesting candidate is alumina-coated silica. It has been described as the 1050 product coated with the 8676 product.

Table 2.5 contains a summary of treatment impacts for several cases considered in this investigation. From this summary, it is clear that the alumina products showed the most promise for future work. In par-

TABLE 2.4. Candidate for Treating HCP Used in This Study.

Candidate	Particle Size	Charge
Colloidal silica (1050)	20 nm	Negative
Sodium silicate (OxyChem50)	Broad range	Negative
Rhoplex	60 nm	Negative
Colloidal alumina (8676)	2 nm	Positive
Alumina-coated silica (1056)	24 nm	Positive

TABLE 2.5. Treatment Permeability Impact and Resistivity Changes.

Candidate	Average Initial Permeability K_o (m/s)	Average Final Permeability K (m/s)	Average Permeability Impact Factor	Average Resistivity Change %
Rhoplex	5E-9	1E-9	8	+4
Rhoplex + $CaSO_4$	4E-9	9E-10	8	+10
Colloidal silica (1050)	8E-9	2E-9	23	+12
1050 + $CaSO_4$	1E-8	2E-9	7	+15
Sodium silicate	2E-8	1E-8	2	+15
Colloidal alumina (8676)	2E-8	5E-9	6	−8
8676 + $CaSO_4$	1E-8	2E-10	95	+27
Alumina coated Silica (A–C–S)	3E-8	1E-9	376	+36
A–C–S+ $CaSO_4$	4E-8	2E-11	21,000	+38

ticular, alumina-coated silica combined with calcium post-treatment exceeded the target permeability reduction. The percent change in resistivity that consistently provided a significant drop in permeability was 27%. In prior work conducted against an opposing seepage flow, the figure was 23%. This is a remarkably close comparison, considering the difference in time of treatment application, which is 5 min vs. 30 min. More time would have provided more opportunity for circuit polarization and a corresponding rise in circuit resistance. In addition, the lack of an opposing flow meant that a source of depolarization was missing from the static results of Table 2.5. This again could have caused the change in resistivity to be higher than in the static cases.

2.2.1. Resistance Changes vs. Permeability Impacts

In general, absolute measures of bulk resistivity are a little cumbersome, but simple changes in resistivity are much easier to obtain. For this reason, the resistivity ratio obtained during treatment was used as a tool for projecting the effectiveness of an electrokinetic nanoparticle treatment. In Figure 2.9, the resistivity ratio is plotted on a linear scale while the permeability ratio is plotted on a log scale. The result is a line that is slightly curved. A reasonable curve kit was obtained using a power law. The base line coefficient of permeability for this curve is $\sim 10^{-8}$ m/s. This was achieved using hardened cement paste specimens with w/c ratios of 0.8. The permeability limit used by the Army Corps of Engineers was, at the time of this research, just above 10^{-11} m/s. The theoretical limit line applies to a coefficient of permeability of 10^{-15} m/s, for a system of capillary pores that has just begun to be discontinuous.

The addition of calcium sulfate treatment exhibited mixed results. Lack of enhancement for the Rhoplex case was not surprising, because no significant chemical reaction was anticipated. The colloidal silica result was surprising, because the expectation was the creation of C–S–H. It seemed that additional trials would, perhaps, tease out a manifest difference, especially considering the large amount of scatter with this work. The largest surprise of all was the unaided impact of alumina-coated silica, which produced a permeability reduction factor of close to 300. Obtaining another 2 orders of magnitude in benefit due to the addition of the sulfate treatments was similarly impressive.

2.2.2. Radical Permeability Reduction Summary

Alumina-containing nanoparticles showed the most impressive performance and the most promise for future applications. In particular, alumina-coated silica combined with calcium post-treatment achieved an astounding 4th order reduction in permeability. The use of circuit

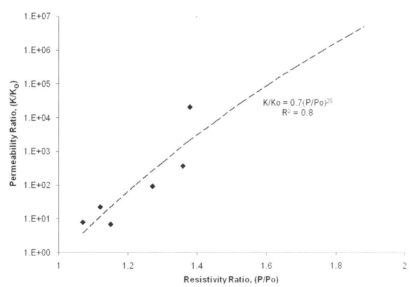

FIGURE 2.9. *Permeability reduction as indicated by resistivity reduction induced by electrokinetic nanoparticle treatment.*

resistance changes will prove to be very useful in determining if a given treatment is considered complete. It is conceivable that a 30 minute treatment is much longer than is necessary to achieve performance that exceeds, for example, the needs of the Army Corps of Engineers.

2.3. REFERENCES

Bazant, Z., Sener, S. and Kim, J. (1987) "Effect of Cracking on Drying Permeability and Diffusivity of Concrete," Technical Paper No. 84-M35, *ACI Materials Journal,* pp. 351–357, September/October.

Brown, P. W., Shi, D. and Skalny, J. (1991) "Porosity/Permeability Relationships," *Materials Science of Concrete II,* The American Ceramic Society, Westerville, OH, pp. 83–109.

Cardenas, H. (2002) "Investigation of Reactive Electrokinetic Processes for Permeability Reduction in Hardened Cement Paste," Ph.D. Thesis, University of Illinois, Urbana-Champaign, September.

Cardenas, H., and Struble, L., "Electrokinetic Nanoparticle Treatment of Hardened Cement Paste for Reduction in Permeability" *American Society of Civil Engineers—Journal of Materials in Civil Engineering,* Vol. 18, No. 4, July/August 2006.

Cardenas, H., Struble, L., Modeling of Permeability Reduction in HCP via Electrokinetic Nanoparticle Treatment, *American Society of Civil Engineers—Journal of Materials in Civil Engineering,* Vol. 20, No. 11, Nov. 2008.

Darcy, H. (1956) *Les fontaines publiques de la ville de Dijon,* Appendix: "Determination of the Law of the Flow of Water through Sand," Victor Dalmont, Paris, pp. 590–594.

Fan, Y., Yin, S., Wen, Z. and Zhong, J. (1999) "Activation of Fly Ash and its Effects on Cement Properties," *Cement and Concrete Research,* Vol. 29, No. 4, April, pp. 467–472.

Finney, D. (1998) "Electro-osmosis Dries Wet Areas," *Cutting Edge,* USACERL, Champaign, IL.

Gratwick, R. T. (1974) *Dampness in Buildings,* John Wiley & Sons, New York.

Hearn, N., Hooton, R. D. and Mills, R. H. (1994) "Pore Structure and Permeability," *STP 169C Significance of Tests and Properties of Concrete and Concrete-Making Materials,* Chapter 25, ASTM, West Conshohocken, PA.

Hock, V., McInerney, M. K. and Kirstein, E. (1998) "Demonstration of Electro-Osmotic Pulse Technology for Groundwater Intrusion Control in Concrete Structures," *U.S. Army Facilities Engineering Applications Program Technical Report 98/86,* p.19.

Hooton, R. D. (1988) "Problems Inherent in Permeability Measurement," *Engineering Foundation Conference on Advances in Cement Manufacture and Use,* Engineering Foundation, August, pp. 143–155.

Kasselouri, V., Kouloumbi, N. and Thomopoulos, Th. (2001) "Performance of Silica Fume-Calcium Hydroxide Mixture as a Repair Material," *Cement and Concrete Composites,* Vol. 23, No. 1, pp. 103–110.

Lim, C., Gowripalan, N. and Sirivivatnanon, V. (2000) "Microcracking and Chloride Permeability of Concrete Under Uniaxial Compression," *Cement and Concrete Composites,* Vol. 22, No. 5, pp. 353–360.

Ludirdja, D., Berger, R. L. and Young, J. F. (1989) "Simple Method for Measuring Water Permeability of Concrete," *ACI Materials Journal,* September–October, Vol. 86, No. 5, pp. 433–439.

Mindess, S., Young, J. F. and Darwin, D. (2003) *Concrete 2nd Ed.,* Prentice-Hall Inc., Pearson Education, Inc., Upper Saddle River, New Jersey.

Ortlepp, W. (1992) "Electroosmotic Process for Coating and/or Impregnating Nonmetallic, Nonconductive Porous Materials with an Inorganic and/or Organic Binder," German Patent CA Section 42 (Coatings, Inks, and Related Products) DE 4113942 A1.

Otsuki, N., Hisada, M., Ryu, J. and Banshoya, E. (1999) "Rehabilitation of Concrete Cracks by Electrodeposition," *Concrete International,* March, pp. 58–63.

Paya, J., Borrachero, M., Monzo, J., Peris-Mora, E. and Amahjour, F. (2001) "Enhanced Conductivity Measurement Techniques of Evaluation of Fly Ash Pozzolanic Activity," *Cement and Concrete Research,* January, Vol. 31, No. 1, pp. 41–49.

Ryu, J. and Otsuki, N. (2002) "Crack Closure of Reinforced Concrete by Electrodeposition Technique," *Cement and Concrete Research,* January Vol. 32, No. 1, pp. 159–164.

Samaha, H. R. and Hover, K. C. (1992) "Influence of Microcracking on the Mass Transport Properties of Concrete," technical paper, *ACI Materials Journal,* July–August, Vol. 89, No. 4, pp. 416–424.

Shi, C. and Day, R. (2000) "Pozzolanic Reaction in the Presence of Chemical Activators, Part II: Reaction Products and Mechanism," *Cement and Concrete Research,* April, Vol. 30, No. 4, pp. 607–613.

Taylor, H. F. W. (1997) *Cement Chemistry, 2nd Edition,* Thomas Telford Pub., London.

Uchikawa, H. (1986) "Effect of Blending Components on Hydration and Structure Formation," *Theme III-Blended and Special Cements, 8th International Congress on the Chemistry of Cement,* Rio de Janeiro, Vol. 1, pp. 249–280.

Vandex, (1983) "Water Proofing Systems," *Vandex product brochure,* Stamford, CT.

Xypex. (1983) "Concrete Waterproofing by Crystallization," *Data Sheet 005,* Richmond B.C., Canada.

CHAPTER 3

Porosity Reduction and Strength Increase

BECAUSE strength is quickly and easily determined from a core sample or a cast specimen, it is a commonly used surrogate for indicating the porosity and thus the durability of concrete. But because porosity largely governs strength as well as long term durability, the two properties are, essentially, inseparable. As porosity diminishes, strength increases. Achieving such changes with nanoparticle treatments is largely a function of particle size, dosage and treatment time. The following sections will present specific techniques and the porosity and strength outcomes that they produce.

3.1. POROSITY REDUCTION

The most direct result of any nanoparticle treatment is porosity reduction. It is the key to both strength enhancement and durability improvement. The amount of porosity reduction achieved depends on how the treatment is conducted. Treating a bulk material tends to achieve a modest result, while targeting an embedded electrode improves the outcome. This enhanced outcome will mostly impact the direct vicinity of the electrode. The following section takes up the more general case of a bulk material treatment.

3.1.1. Bulk Material Porosity

When a nanoparticle treatment is distributed throughout a bulk material, the porosity reduction is somewhat diffuse because it depends upon randomized incidences of bottlenecking, coagulation, and the settling of particles throughout the porous network. This broad distribution of particles is achieved whenever the destination electrode is very large or broadly distributed; it is also the case if the embedded electrodes are a large distance away from the starting point of the particles, because this prevents them from arriving prior to the completion of treatment. Rather than concentrating at a specific destination point, they are distributed to wherever the electric field lines cause them to spread.

The porosity reduction findings listed in Table 3.1 were taken from cylindrical specimens exposed to damaging levels of sodium sulfate for 1 month (Cardenas *et al.* July, 2011). The centers of the 3 inch diameter specimens were batched with a 1/16-inch titanium wire that was used to electrokinetically draw nanoparticles toward the center of each specimen. The treatment was dosed at approximately half the specimen storage capacity, and the treatment duration was only 5 days at a low voltage insufficient for drawing particles all the way to the titanium wire.

The result was a distributed nanoparticle treatment. Table 3.1 lists the 3 specimen cases in terms of the type of exposure. The lime-water exposure provided for an undamaged control category for comparison, starting with compressive strength and including two parameters obtained from mercury intrusion porosimetry (MIP). The sulfate-exposed specimens exhibited gross spallation and a 30% porosity increase. The sample treated with electrokinetic nanoparticles (EN) exhibited a return to the pre-damaged porosity level.

The threshold pore diameter is generally interpreted as the pore-size associated with initial access to the interior of the material. In the case of the specimens exposed to sulfate this diameter nearly doubled in size; but when subjected to bulk material nanoparticle treatment, their threshold pore diameter dropped dramatically, from 220 μm down to 58

TABLE 3.1. Response of Sulfate-Damaged Concrete to Bulk Material Nanoparticle Treatment.

Type of Exposure	Compressive Strength (psi)	Porosity by MIP* (%)	Threshold Pore Diameter (μm)
Lime water	4000	31	123
Sulfate Exposed	2100	40	221
EN Treated	2800	29	58

*MIP = mercury intrusion porosimetry.

μm. This remarkable change speaks to the capacity of nanoparticles to form pile-ups throughout the pore system at various bottlenecks, even at points of initial access. A diffuse and distributed particle treatment evidently exhibits this random pile-up tendency to which MIP analysis is very sensitive.

The impact on porosity was far less dramatic. Dropping from a 40% porosity to 29% constitutes a porosity reduction of 28%. This is a typical result when nanoparticles are diffusely distributed over a bulk material volume. Despite such modest reduction, the material strength responded with a 33% increase—which was somewhat lower than anticipated, given that this type of treatment more often yields a strength increase of approximately 50%. The reason is clear. As noted earlier, macroscopic spallation damage was observed on the exterior walls of the sulfate-damaged specimens. This level of damage cannot be impacted by nanoparticle treatment, because the defects are too large. Clearly, the only strength-enhancement achieved was through penetration of smaller cracks and pores throughout the bulk material.

3.1.2. Electrode Packing Treatment

When nanoparticles are permitted to travel to an embedded electrode, they become more densely concentrated than when they are diffusely distributed through the bulk material. Figure 1.3 illustrates how this is believed to work with a composite particle such as alumina-coated silica. When the 20-nm silica particles are impeded due to a barrier such as the steel reinforcement, the electric field can pull the 2-nm alumina particles from the silica surface and send them on to the reinforcement surface. This enables the nanoparticles to pack densely the capillary pores in the vicinity of the embedded electrode. The porosity reduction in the vicinity of this electrode is greater than that observed in the bulk material some distance away. The bulk material does still benefit from random deposits of nanoparticles; but the material next to the embedded electrode is much more enhanced, because nanoparticles from a broad area are being funneled toward it.

Table 3.2 lists parameters of durability for reinforced concrete cylinders that received nanoparticle treatment prior to a year-long exposure to simulated seawater (Cardenas *et al.* June, 2011). The parameters were measured after the exposure period. Each cylinder (3 inches in diameter) contained at its center a length of 1/4-inch diameter, 1018 steel rod. The treatment lasted 2 weeks and was designed to pile up

TABLE 3.2. Concrete Porosity, pH, and Corrosion Rates at a Reinforcement that has Received Electrokinetic Nanoparticle (EN) Treatment.

Treatment Type	Porosity (%)	pH	Corrosion Rate (mpy)
Control	9	8.4	37
EN-Treated	5	11.8	0.5

all the alumina-coated silica nanoparticles at the steel reinforcement. The parameters listed in Table 3.2 were measured after 1 year of post-treatment exposure to water-line saltwater immersion.

After a year of simulated seawater exposure, the porosity of the control samples was 9%. The porosity of the EN-treated specimens was reduced by 44% to a value of 5%. The pH of the concrete adjacent to the steel reinforcement provided an indication of the severity of this environment. The naturally high pH in concrete protects iron from corrosion as long as the chloride content is not too high. Fresh concrete can exhibit a pH in the pore fluid as high as 13.8. After a year of waterline exposure to simulated seawater, the control specimens had a pH of only 8.4 and the elevated corrosion rate reflected this problem. A corrosion rate above 1.0 mils per year (MPY) is considered a high value for most structures. In contrast, the EN-treated specimens exhibited a relatively good pH of 11.8. The lower corrosion rate of 0.5 MPY confirmed the improved durability of the nanoparticle treated materials.

Focusing the treatment directly onto the reinforcement caused a dense pile up of nanoparticles in the pores adjacent to the rod. The small particles formed a dense barrier, in which the interparticle spacing was small enough to keep most of the chlorides out. This barrier also appeared to protect the environment adjacent to the electrode from losing alkali ions due to leaching. Low leaching rates permit the pH of the pore fluid in this vicinity to remain at a high level, which keeps the iron in the passive state.

3.1.3. Long Term Electrode Packing

Another key influence on porosity reduction is the duration of treatment. Longer lasting treatments deposit species from the pore-fluid environment that further densify the pore system. In a typical electrochemical chloride extraction process, the treatment period may last 6–8 weeks. The treatment fluid will typically contain one or more re-alkalizing agents. These species, plus what is available in the existing pore fluid, will form some of the deposited material at the reinforcement. Some level of dissolution may also take place along the pore walls, leading to more ions being deposited at the reinforcement. This by itself, without the addition of nanoparticles, can lead to significant porosity reduction in the reinforcement's vicinity. The advantage of using nanoparticles stems from the added mass-action event. Solid particles are simply a more time-efficient way to move significant quantities of matter. Later sections will examine how these time efficiencies increase substantially as the applied current density of the treatment increases. For now, the examination will focus on packing dynamics as a function of treatment time and dosage.

The data presented in Table 3.3 relates to a series of 2-m long reinforced concrete beams that were cast for long term corrosion work. The beams were cast with # 4 and # 5 iron rebars and a 4.5 wt% content of NaCl in the mix water. Nanoparticle treatments conducted with alumina-coated silica were dosed in terms of the liters of nanoparticles being delivered per unit of concrete surface area adjacent to the reinforcement. Each dosage was calculated to yield a specific thickness of nanoparticle deposit. For example, the 0.16 liter/m^2 dosage was designed to provide a 1/2-inch layer of particle-densified pores surrounding each element of the reinforcement. Each of the other dosages represented a doubling of this deposited layer of protection. This means that 0.33 l/m^2 correlated to a 1-inch design layer and the 0.65 l/m^2 dosage corresponded to a 2-inch layer of protection. These dosages were applied over a period of 6 weeks, immediately following a 2-week electrochemical chloride

TABLE 3.3. Pore Structure Impacts of an 8 Week Electrokinetic Nanoparticle Treatment.

Treatment Case/Dosage	Threshold Pore Diameter (μm)	Porosity (%)	Porosity Reduction (%)
Control	773	15.3	—
ECE	482	10.4	32
0.16 (l/m^2)*	352	8.4	45
0.33 (l/m^2)	294	4.9	68
0.65 (l/m^2)	64	2.8	82

*Dosage listed as liters of particles per square meter of treated surface.

extraction treatment. The pore structure data of Table 3.3 was obtained after the beams were partially submerged in simulated seawater for 6 months and then destructively tested.

Mercury intrusion porosimetry was used to examine the pore structures. The samples were harvested approximately 1.5 to 2 inches away from the reinforcement. Some of the particle dosages would be more effective than others at this distance. The untreated control specimens exhibited a threshold pore-diameter (TPD) of 773 μm and a total volume porosity of 15%. The electrochemical chloride extraction process (ECE) produced a TPD of 482 μm and a porosity of 10%. This clearly provided a distribution of deposits from the alkali metals in the treatment fluid as well as ions originating from within the cement pore fluid. By comparison, the lowest nanoparticle dosage of 0.16 l/m^2 provided a TPD of 352 μm and volume porosity of 8.4%, which constitutes a reduction in porosity of 45%.

Further increases in nanoparticle dosage provided for continued and remarkable improvement in porosity reduction. The highest dosage provided a porosity of only 2.8% for an 82% porosity reduction. The TPD was reduced even more profoundly by a factor of 12. The reduction in porosity was certainly beneficial as a result of the ECE process. The outcomes were remarkably improved as a result of nanoparticle addition to the treatment fluid.

It is interesting to compare these results to the void space present at the atomic scale. In these circumstances, the amount of matter present in the unit cell of an atomic crystal structure can be as high as 86%. Achieving a porosity reduction above 50% usually means that additional ion species are part of the deposit. The finely-placed deposits that are achieved by these ionic species are able to fill the interparticle spacing within the packed structures of the deposited nanoparticles. Such extreme porosity reductions are achieved after long-term treatments, but could be accelerated if suitable types and numbers of ionic species are also made available. Care must be taken to keep the nanoparticle suspensions from becoming unstable as a result of any added ionic species, as well as those that may be drawn from the concrete pores and into the suspension fluid.

3.1.4. Porosity vs. Permeability

How to distinguish easily between porosity reduction and permeability reduction is not exactly clear. The real difference comes down to the connectivity and tortuosity of the pore network. In theory, two different mixes, while exhibiting the same porosity and strength, may display quite different permeabilities due to the degree of continuity in their respective pore networks. In practice, the two will change together, since the randomized processes that reduce porosity will reduce connectivity at the same time. A key difference between reductions in porosity and permeability is that a great deal of permeability reduction can be achieved at or near the surface without having to conduct a bulk section treatment. This approach saves time and materials costs. Distinguishing between the two may not be terribly important as long as one understands that the benefits of reducing permeability can be achieved at the surface alone, while the increase in strength from porosity reduction must happen throughout the bulk of the material.

3.1.5. Porosity Reduction in Summary

As noted at the beginning of this section, porosity reduction is dependent upon how broadly the particles are allowed to distribute over the volume of concrete. When bulk material is treated, porosity reduction for a short-term treatment (2 weeks or less) will be on the order of 30%. If particles are driven toward an embedded electrode (such as rebar), the local reduction in porosity for short-term treatment will be on the order of 50%. If the treatment is properly dosed (a design topic covered later) and is permitted to last a couple of months, then ions originating from the concrete matrix (or elsewhere) can contribute to achieving porosity reductions that reach 70–80%.

3.2. STRENGTH ENHANCEMENT

Increase in strength results primarily from reductions in porosity. This means that treatments designed for strength enhancement will be similar to those used to obtain porosity reduction. Another means of increasing strength is by providing nanoparticles or ion species that react to form stronger phases within the pore, or by converting weak phases into stronger ones. Some important restrictions and considerations apply to strength gain. These considerations will be described in the following sections.

3.2.1. Strength Over Time and Maturity

After concrete is batched, its porosity and strength change over time. Hydration reactions continue to decrease porosity, which correlates to increase in strength. When concrete is far along in the process of gaining strength, it indicates that the hydration reactions are largely complete. The concrete is then considered matured.

Table 3.4 displays data obtained from cylindrical specimens treated with a 24-nm alumina-coated silica particle for corrosion control of the steel reinforcement. The 1/4-inch diameter 1018 steel rods were centrally located along the length of the 3-inch diameter specimens. Maximum aggregate size was 3/8 inch. Nanoparticles entered through the sides of the specimens and were concentrated at the steel rods. Each specimen was treated for 7 days with a current density of 1 A/m^2 as calculated over the outer surface of the concrete. Half the batch was allowed to soak in simulated seawater for 1 year prior to nanoparticle treatment. The other half was treated 28 days after batching. In each case the specimens were soaked in simulated seawater for an additional year prior to evaluation for corrosion and indirect tensile strength. The embedded steel acted as a long void in the center of each specimen. Since this influence was the same for all cases, the comparisons for age and treatment were clearly distinguishable.

It was found that mature specimens exhibited a better response to treatment in terms of tensile strength, which increased by over 100%. By comparison, young specimens only experienced a gain in strength of slightly less than 20%. The high corrosion rate experienced by control specimens had probably deteriorated the strength long before nanoparticle treatments took place. These treatments radically reduced the corrosion rate in the mature concrete by nearly a factor of 10, and yielded a tensile strength almost equivalent to the finishing strengths of the young

TABLE 3.4. Strength vs. Maturity at Time of Treatment.

Treatment Type	Specimen Age at Treatment (months)	Corrosion Rate 1 Year after Treatment (mpy)	Strength 1 Year after Treatment (psi)	Strength Increase (%)
Mature control	12	517	263	—
Mature EN-treated	12	66	535	103
Young control	1	37	583	—
Young EN-treated	1	0.5	680	17

concrete. This data touches on a time-dependent restoration theme. It appears that the more advanced deterioration was more profoundly rehabilitated by the nanoparticle treatment. As noted, however, in the case of sulfate damage that had caused irrecoverable spallation, the limit of this outstanding rehabilitation capability probably has to do with cracks becoming so large that nanoparticle treatments are insufficient for the scale of the restoration.

There is another time-dependence issue that can impact the magnitude of strength gain. The young specimens received particle treatments into a pore system that was still changing. The control specimens had pore structures that continued reducing in porosity and connectivity as hydration proceeded. These ongoing changes to the pore structure of the controls appeared to overshadow much of the gain in strength that the nanoparticles had provided to the companion specimens, which is probably why the difference in strengths between the cases was only 17%. This is reminiscent of how silica fume increases early strength when it is used in a given concrete mix. Similar early strength enhancements take place when nanoparticles are used directly in a concrete mix. At the time of treatment, the control specimens of the mature concrete were not experiencing significant ongoing changes to the pore structure because the hydration reactions were largely complete, thus the changes induced by the nanoparticles were manifest in a much more static system. The increase provided by the nanoparticles (103%) had a much better chance of remaining significant because there was little ongoing hydration process left to overshadow its impact.

3.2.2. Strength Gain and Phase Creation

When a structure presents two parallel surfaces accessible to treatment, as in masonry block walls, the opportunity exists to deliver reactive agents from each surface and have them meet and react within the interior. One benefit of this is the ability to introduce new phases that cannot be transported but can be formed reactively where the reaction products meet in the pores. Another is that the porosity reduction achieved is generally greater than that of a single phase treatment distributed throughout the bulk material. This may be due to the fact that two oppositely charged reactants meeting in the middle tend to fit through smaller pores while in transit, and their collision tends to create a more consolidated phase. It is as if they were arriving at an embedded electrode.

TABLE 3.5. Compressive Load Test Results for Masonry Blocks.

Block Weight Cases	Failure Load * (lbs)	Increase in Strength* (%)	Block Mass (kg)
Light-weight control	8900	—	12
Light-weight treated	20,600	130	12.47
Heavy-weight control	14,800	—	14.4
Heavy-weight treated	30,500	105	14.9

A masonry block treatment can be arranged by providing sodium silicate in the interior and a calcium hydroxide solution along the exterior. The calcium ions meet the silicate ions along the interior pores and form a variant of calcium silicate hydrate. The findings listed in Table 3.5 relate to such a treatment conducted for 5 days with a current density of 9 mA/m² of outer surface area.

Each weight class of block responded well to treatment. The light-weight block strength gained 130% in strength. The heavy-weight block gained 105% in strength. In any case, both of these gains in strength were remarkable in that they occurred in a period of only 5 days. Figure 3.1 contains images of two of the fracture surfaces. The treated block showed areas of a whitish deposit that is indicative of the electrokinetic treatment.

The two weight classes of block differed in mass by approximately 20%. The light-weight block exhibited a starting resistance of 8900 lbs, the heavy-weight block was 66% stronger with a failure load of 14,800 lbs. The 5-day treatment when applied to the light-weight block caused it to exceed the strength of the heavy-weight block by 40%, while ending at a mass that was still 15% lighter. This provided a specific strength improvement of 30%.

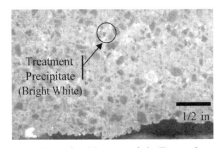

Reactive Nanoparticle Treated   Untreated Control

FIGURE 3.1. Fracture surfaces of light-weight masonry block following compressive load tests.

3.3. REFERENCES

Cardenas, H., Kupwade-Patil, K., Eklund, S., "Corrosion Mitigation in Mature Reinforced Concrete using Nanoscale Pozzolan Deposition", *American Society of Civil Engineers—Journal of Materials in Civil Engineering,* Vol. 23, No. 6, 13 June, 2011.

Cardenas, H., Kupwade-Patil, K., Eklund, S., "Recovery from Sulfate Attack in Concrete via Electrokinetic Nanoparticle Treatment", *American Society of Civil Engineers—Journal of Materials in Civil Engineering,* Vol. 23, No.7, July, 2011.

CHAPTER 4

Crack Repair

C RACKING in concrete is a continuing nuisance that makes trouble on many levels. It is caused by drying shrinkage, foundation problems, corrosion problems, and other modes of deterioration. The biggest question raised by any given crack is what it looks like in locations that cannot be directly viewed. Though relatively thin on the surface, it may be much larger in the interior. To reveal these unknowns, the crack is opened further and a non-shrinking grout is applied to patch the area. Unfortunately, the patching operation results in a "cold joint" between the old concrete and the new patch, with reliability well below that of either the base material or the patch cement. This joint is then the most likely place for a new crack or leak to occur. The next sections describe nanoparticle treatment observations and operating principles that lead to a cold joint performance greatly exceeding our dismal expectations.

As we have already observed, nanoparticles tend to pack tightly around an electrode embedded in concrete. In this packed region, the strength of concrete is expected to rise. Figures 4.1 and 4.2 illustrate the experimental setups that were used to explore this idea and to achieve radically enhanced crack repair. This exploration was undertaken in the Applied Electrokinetics Laboratory under the direction of the author at Louisiana Tech University. Joshua Alexander, Israel Popoola, and Mark Castay provided extensive assistance. In Figure 4.1 the repair of a cracked beam was accomplished by removing unsound material and opening a location to apply cementitious grout to a modified crack of known dimensions. Strength evaluation of the repair was conducted via third-point bending to determine the modulus of rupture (ASTM C78). This test is designed to evaluate how much of a bending loading the

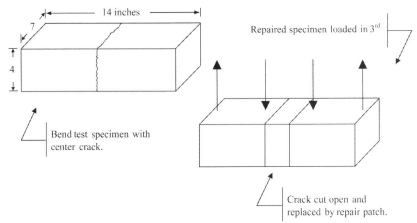

FIGURE 4.1. Beam specimen design used for experimental crack repair work.

material can withstand prior failure. In virtually all cases, the weakness of the patch/base material interface led to failure at the interface. This interface is effectively the most porous location in the repair. A future failure in the repair will typically occur at this interface, because the interlock between the two materials is not as secure as the grain-on-grain interlock existing in the base material.

4.1. GETTING NANOPARTICLES TO A CRACK REPAIR

To enhance the strength and reduce the permeability of the repair interface, we borrowed methods from previous corrosion repair work by

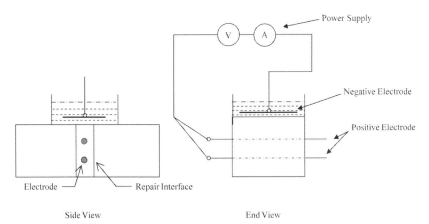

FIGURE 4.2. Beam treatment setup with wire mounted away from the repair interface.

casting a narrow wire into the repair patch. Application of an electrokinetic treatment drove nanoparticles toward the embedded wires shown in Figure 4.2. The arrival of the particles caused the porosity in these regions to decrease, yielding a significant permeability reduction and an increase in strength.

The vessel mounted above the beam carried a fluid of nanoparticles suspended in liquid. An electrode immersed in the liquid was positively charged and used to drive the particles into the pores of the concrete and toward the embedded wire electrodes. During this transport process, the expectation was that some of the particles would be caught and held at various locations, causing general strengthening throughout the concrete. As particles hit the wire, additional particles would join the collection and thus each adjacent pore would become impacted with a dense collection of particles. This higher level of strengthening would be expected to start at the electrode and then propagate back along each pore. The intention was that this back-up of particles would densify the region up to and including the repair interface.

There are three types of specimens presented in Table 4.1: a specimen that was repaired and electrokinetically treated; one that was only repaired; and one that was not damaged or repaired in any way. The treated repair exhibited a modulus of rupture of 400 psi. Typical compressive strength could be anywhere from 4000 to 10,000 psi. The low strength-level of a rupture modulus is due to the mode of loading. This load causes tensile stresses to develop at the bottom of the beam. Failure is thus expected to be controlled by the tensile strength of the material, which is generally a factor of 10 lower than the compressive strength.

TABLE 4.1. Modulus of Rupture for Beams with Wires Mounted Off the Repair Interface.

	Specimen	Fracture Load, P (lb)	Strength (psi)	Ave. Strength (psi)
Treated repair	1	2800	480	
	2	2700	394	400
	3	2650	387	
Untreated repair	1	1000	146	
	2	1200	175	160
	3	1150	168	
Undamaged	1	3820	557	
	2	3860	563	560
	3	3840	560	

By comparison, the untreated repair exhibited a modulus of 160 psi. This means that the electrokinetic treatment caused the crack repair to be 2.5 times stronger than previously possible. Such an improvement is expected to enhance the durability of the structure as compared to a conventional repair.

 It is convenient to describe the impact of a treatment on a repair in terms of the following factor:

$$\text{Strength Increase Factor (SIF)} = \frac{\text{Treated Repair}}{\text{Untreated Repair}} = 2.5$$

The SIF in this case indicated that the treatment improved the repair interface strength by a factor of 2.5. When considering the impact of damage that was conventionally repaired, one could use another factor to characterize a reduction in strength as follows:

$$\text{Strength Reduction Factor (SDF)} = \frac{\text{Undamaged}}{\text{Untreated Repair}} = 3.5$$

In this case, the factor indicates that the conventional repair yielded a specimen whose repair interface was weaker than the original base material by a factor of 3.5. The treated repair improved this strength radically by a factor of 2.5.

Later work in this study examined the opportunity to enhance the strength of the repair even further by locating the wires not in the center of the repair patch, but rather directly at the repair interface. This approach was expected to concentrate more particles at the critical location.

4.2. SIMULATING THE PARTICLE PACKING PROCESS FOR CRACK REPAIR

Before this treatment enhancement is examined further, it is valuable to first consider the results of graphic simulation modeling conducted to determine the potential impact of an electrokinetic treatment. A key aim of such treatment is net reduction in porosity, which is directly connected to both the strength and the permeability of concrete. Permeability is a vital consideration when one seeks to extend the service life of a concrete structure. The less permeable the concrete, the fewer aggressive chemicals can gain access to it. Carbon dioxide, de-icing salt, sulfates and other substances can cause both short and long term reduction in strength. Figure 4.3 shows an image obtained while the simulation was operating. In this case, the pore is represented by a cylinder with walls consisting of linear, truss-like members. The attraction of the negatively charged wire draws the simulated particles toward the bottom of the cylinder. The blue spheres represent 10 nm particles of silica. The white spheres represent 2 nm particles of alumina. The black spheres represent 0.56 nm diameter assemblies of solvated chloride ions.

In this work, several simulations were run in order to examine how the porosity of concrete would be impacted one cylindrical pore at a time by alumina-coated silica nanoparticle assemblies. Figure 4.4 illustrates some of the findings, focusing on selected combinations of particle size and pore diameters ranging from 30 to 40 nm.

The case involving 2 nm and 4 nm (2 + 4) particles is illustrated by diamond shaped points in Figure 4.4. For 20 nm pore size, the porosity reduction was approximately 48%. For 30 nm pore size, the porosity reduction improved to just over 50%. The result for the 40 nm pore was also ~50%. In general, the combination of small and large particles appeared to reduce porosity better as the larger of the two particles in-

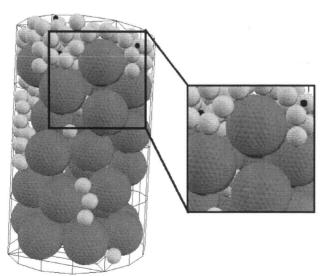

FIGURE 4.3. *A packing simulation output showing 6 nanometer particles (dark) and 2 nanometer particles (white). The solvated chloride ions are black.*

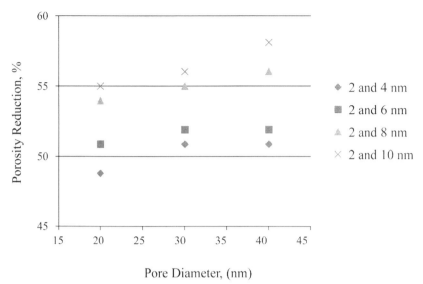

FIGURE 4.4. Simulation data porosity reduction.

creased in size. This makes sense, since it permits more of the pore to be occupied by solid phases as opposed to a very large number of smaller particles (and the large amount of inter-particle spacing that would be formed among them).

The best porosity reductions observed in this study were just under 60%. This is well below the theoretical limit of 86% observed among densely packed atoms. The difference between the simulation result and the theoretical limit stems from the way the boundary for each case is conceived. In the case of the current study, the particles are not permitted to go beyond the walls of the cylindrical pore model. Consequently the packing density of the particles adjacent to the pore walls was not as high as those located toward the center of the pore, and the porosity reduction was well below the theoretical limit.

It is notable that typical experimental treatments tend to achieve porosity reductions in the vicinity of 50% for a 2 and 20 nm particle-size combination. This amount of porosity reduction was typically confined to the areas directly adjacent to the electrode within the concrete. Further away from the electrode, this reduction in porosity was less effective. Such loss in effectiveness was due to the fact that the particles were not able to thoroughly densify all regions away from the electrode. It is also notable that 50% porosity reduction is the lowest value obtained in the simulation work shown in Figure 4.4.

4.3. PLACING THE PARTICLE PACKING ELECTRODE

Based on these simulations and related work, it became apparent that moving the embedded electrode closer to the repair interface may provide an enhanced repair. By placing the embedded wire directly on the repair interface, the densification due to treatment could be radically concentrated there, thus providing a targeted enhancement to this weak location. Figure 4.5 illustrates the wire electrode placement (one at each interface). This electrode was placed relatively close to the top surface of the specimen in order to speed up the treatment transport rate, and because of the high tensile stresses anticipated at this surface when a bending load was applied.

Several tests were conducted using the configuration of Figure 4.5. Table 4.2 contains the results of several of these tests. It was observed that the treated repair specimens exhibited a modulus of rupture of 210 psi. The untreated repair provided a higher value of 250 psi. This compares unfavorably to the undamaged beam, which provided an average value of 960 psi.

Clearly the treatment did not appear to enhance the rupture modulus of the repaired beams. The strength increase factor (SIF) had a value of only 0.83. By comparison, the SRF was 3.8. In order for the treatment to be successful an SIF of greater than 1 and preferably greater than 3.8 would be needed.

After attempting various permutations involving longer treatments and other particle candidates, it became evident that the location of the wire in Figure 4.5 could be a source of difficulty. From prior work it was observed that the strength of a given concrete specimen behaved as if

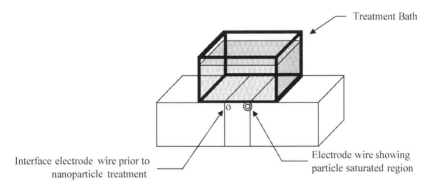

Treatment Bath

Interface electrode wire prior to nanoparticle treatment

Electrode wire showing particle saturated region

[Bottom electrodes run entire length of interface]

FIGURE 4.5. Beam treatment setup with wire mounted at the repair interface.

TABLE 4.2. Modulus of Rupture for Beams with Wire Mounted at the Repair Interface.

	Specimen	Fracture Load, P (lb)	Strength (psi)	Ave. Strength (psi)
Treated repair	1	2200	320	
	2	1400	200	210
	3	1000	150	
	4	1200	180	
Untreated repair	1	2200	320	
	2	—	—	250
	3	1600	230	
	4	1400	200	
Undamaged	1	6680	970	
	2	7154	1040	960
	3	5900	859	
	4	6800	990	

SIF = 210/250 = 0.84. SRF = 960/250 = 3.8.

the wire represented a hole in the component. Experimental work with cylindrical specimens indicated that treatments provided a strength increase despite the presence of this hole. This hole, however, was never before located at a weak point in the material, such as at a repair interface. It now appears that the location of this hole near a weak point constitutes a severe and unnecessary limitation.

4.4. OPTIMIZING THE PACKING ELECTRODE PLACEMENT

Additional analysis has shed light on how this situation can be resolved. It appears that the repair interface can be viewed as a thin region of high porosity due to a lack of adhesion between the repair and the base material. The issue is: how much of a change in porosity achieved near an electrode can influence this area of low strength? A theoretical expression for the relationship between strength and porosity is provided as follows (Mindess, *et al.* 84–87),

$$\sigma'_c = 34,000 * X^3 \text{ psi} \tag{4.1}$$

where

$$X = \frac{0.68 * \alpha}{0.32 * \alpha + w/c} \tag{4.2}$$

where σ_c' is the compressive strength and X is called the gel space ratio, alpha is the degree of hydration (or curing) and w/c is the ratio of water-to-cement (by mass) in the concrete mix. The X value on its own is not as convenient to work with as the capillary porosity. This factor can be related to the capillary porosity (P_c) by noting that,

$$P_c = \frac{w/c - 0.36 * \alpha}{w/c + 0.32} \tag{4.3}$$

By eliminating alpha from Equations (4.2) and (4.3), the gel space ratio now becomes a function of capillary porosity (which is easy to measure) and the w/c ratio (which is easy to control) as given by,

$$NX = \frac{1.8889 * P_c * w/c + 0.6044 * P_c - 1.8889 * w/c}{0.8889 * P_c * w/c + 0.2844 * P_c - 1.8889 * w/c} \tag{4.4}$$

After substituting the new gel space ratio into Equation (4.1), an equation that relates capillary porosity to the compressive strength of the concrete was developed. This equation was then inserted into a spreadsheet and a curve of the compressive strength of the cement versus capillary porosity was generated. This curve is shown in Figure 4.6. A trend line was calculated which gives a simpler relationship. This trend line is also shown in Figure 4.6, where the x variable is the capillary porosity and the y variable is the compressive strength.

Knowing that the modulus of rupture is approximately one-tenth the capacity of the compressive strength, an equation relating tensile strength to porosity can be represented as follows.

$$\sigma_t = 2,668 * P_c^2 - 9,106 * P_c + 3,400 \text{ psi} \tag{4.5}$$

As noted in the previous tables, the SDF ranged from 3.5–3.8 due to the weakness of the repair interface. It is clear now that when a hole is

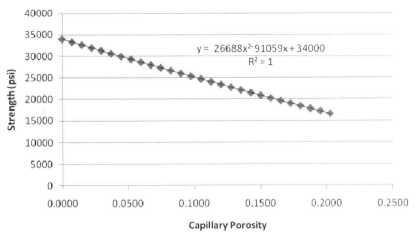

FIGURE 4.6. Relationship between capillary porosity and compressive strength.

located along this interface, the stress concentration effect causes the stress in this location to increase by a factor of 3. This means that the observed rupture modulus could now drop to as low as 3.8 × 3 or ~12 times lower than the undamaged material. The application of a treatment would reduce the capillary porosity, P_c, by as much as ~50% from 0.4 to 0.2. Applying this change in porosity to the strength [Equation (4.5)] would yield a strength increase of as much as a factor of 9. The resultant SIF would become 9/(3.8 × 3) = ~ 0.8. This compares closely to the 0.84 value for SIF observed in Table 4.2, when the wire was placed adjacent to the repair interface.

When the wire was located away from the repair interface, the stress concentration was significantly reduced. The resultant SIF in this case became 9/3.5 = 2.6. This compares relatively closely to the SIF of 2.5, observed in Table 4.2, that was derived from the case in which the wire was not placed in contact with the repair interface. This means that the prediction was quite close.

The key question now was to determine how close the wire electrode can be located to the repair interface and to the peak tensile stress surface of the beam. To answer this question, analytical models governing stress concentration near a hole and another for stresses in a bending beam needed to be utilized. Figure 4.7 defines the physical nature of this stress problem. The lower edge of the rectangle represents the neutral axis of the beam. The top of the beam is the location of highest applied stress. The origin of the x'-y' axis is the location of the wire electrode some distance y' from the top of the beam (highest stress) and some distance x' from the repair interface (weakest location). Placing the wire electrode close to either of these boundaries will enable a highly-densifying treatment to saturate these susceptible regions with a strengthening dose of nanoparticles. Setting the wire electrode too close will permit the stress concentration factor adjacent to the wire to radically increase the stress levels at these vulnerable regions. Fortunately, the magnitude of this localized stress increase drops at a parabolic rate as the wire is moved away from the location of interest.

In order to identify locations within the repair patch at which the strength enhancement is productive, several equations were solved simultaneously. Expressions utilized included the stress field in the vicinity of the wire electrode and the tensile stress due to bending. The influence of repair interface weakness was also included. These expressions were assembled and examined with the objective of determining the locations at which electrode placement would permit effective porosity reduction without a damaging stress concentration. Figure 4.8 illustrates the results in terms of a boundary line separating the zones in which electrode placement could or could not be expected to yield satisfactory benefit. In the satisfactory zone, electrokinetic treatment is expected to provide enhanced repair strength as compared to a conven-

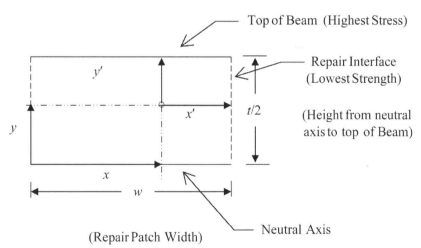

FIGURE 4.7. *Geometric definitions applied to a repair patch set up to provide electrokinetic treatment. The wire electrode is set at the origin of the x'-y' axes. Moving the electrode close to the top of the beam causes the peak tensile stresses to be multiplied by a factor of 3. Setting the wire close to the repair interface will place this factor of 3 stress increase at a location of severe weakness. The need is to find those locations that are close enough to promote significant porosity reduction while preventing the severe stress concentration factor of the wire from cancelling out the treatment benefit.*

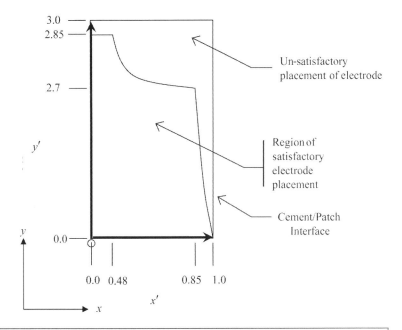

Note: y' is the vertical distance from the wire to the top surface of the beam.
   x' is the horizontal distance from the wire to the cement/patch interface.
   x is the horizontal distance from the wire to the center of the repair patch.
   y is the vertical distance from the wire to the neutral axis of the beam.

FIGURE 4.8. *Diagram showing predicted zone for acceptable wire electrode placement.*

tional repair. The unsatisfactory zone indicates locations in which electrode placement would yield poor mechanical performance due to stress concentrations near the wire, and because of the low strength material present along the repair interface. An interesting question is whether the electrokinetic repair enhancement can be made large enough to completely overcome the strength deficiency of the repair interface. Even the strength enhancement of a factor of 2.6 achieved thus far is still lower than the repair interface depreciation, which can be down in strength by as much as a factor of 3.1–3.8. The enhancement factor of 2.6 was obtained from a 50% porosity reduction achieved via a short term treatment. Lengthening the treatment time could improve this result considerably, as demonstrated in Section 3.1.3, in which the porosity reductions were in the vicinity of 70–80%. It is conceivable that achieving an 80% porosity reduction could be enough to overcome the repair interface depreciation of 3.1–3.8. This could mean that a longer treatment could yield a repair interface region that is actually stronger than the undamaged base material.

4.5. SUMMARY OF ELECTROKINETIC CRACK REPAIR

Conventional crack repair technology yields cold joints that exhibit disappointing mechanical properties and little anticipation of durability. Training nanoparticles to address this deficiency has been a matter of finding the right place to locate an electrode within a repair patch. Getting particles to reduce the porosity adjacent to this electrode means that the porosity and strength at the repair interface is also enhanced. Unfortunately, if this packing electrode is too close to the repair interface, then the stress concentration associated with the electrode defeats much or all of the particle packing benefits. Optimizing the location of the packing electrode within the patch material has been the key to enhancing crack repair. It appears that the day is coming when a crack repair will exceed the strength of the base material.

4.6. REFERENCES

ASTM C78/C78M-10, "Standard Test Method for Flexural Strength of Concrete (Using Simple Beam with Third-Point Loading)," *American Society of Testing and Materials International,* West Conshohocken, PA, 2010.

Mindess, S., Young, J. F. and Darwin, D. (2003) *Concrete 2nd Ed.,* Prentice-Hall Inc., Pearson Education, Inc., Upper Saddle River, New Jersey.

CHAPTER 5

Chloride Extraction and Nanoparticle Barrier Formation

ELECTROKINETICS has been used for the decontamination of soils for some time, and its use in concrete has also been practiced for decades. The application of nanomaterials to these processes now makes it possible not only to decontaminate but also to remedy the circumstances that give chlorides access to reinforcement. The following sections focus on decontamination of concrete containing chlorides, the formation of a protective barrier, and the impact on corrosion resistance. Many of these extraction and barrier development principles are applicable to any other contaminants that exist in ionic form.

5.1. CHLORIDES AND CORROSION

The most destructive and costly contaminant in concrete is the chloride ion. It participates in reactions that cause the dissolution of iron in concrete. Normally, the high pH existing inside concrete keeps the iron stable. Unfortunately, when chlorides diffuse into the pores and make contact with iron, this passive state is compromised. Worse, the chloride ions are not consumed during the corrosion process, but instead are efficiently recycled as they catalyze a destructive transformation of iron metal into iron oxide. This oxide can undergo many changes in stoichiometry which causes the oxidized version of the iron to absorb water and occupy a much larger space. Such expansion causes serious cracking in the surrounding concrete and terribly high costs for repair.

Over the last few decades, corrosion of iron in concrete has become a serious issue both in warm and cold climates, and especially in coastal locations (Bentur, Diamond and Berke, 1998; Broomfield, 1997). It is an $8 billion national repair problem (Koch *et al.*, 2002). One of the causes of corrosion is the diffusion of chlorides, derived either from the use of de-icing salts or exposure to marine environments (Bentur, Diamond and Berke, 1997). Corrosion reduces the effective cross-sectional area via formation of corrosion products. The increased volume that the corrosion products occupy results in tensile cracking that leads to structural failure. Electrochemical chloride extraction (ECE) has been applied to concrete structures for decades with good results.

The existence of various types of positively-charged nanoparticles enables the implementation of the same electric circuits that are used for cathodic protection or chloride extraction to now be used for injecting these protective species into concrete and directly onto the reinforcement. The following sections reveal how this is done, and the exciting advantages over current decontamination practice that is now at hand.

5.2. CHLORIDE REMOVAL FROM STRUCTURES

By applying a constant current for a duration of 1.5–2 months, ECE drives the removal of chloride ions from concrete (Marcotte, Hansson and Hope, 1999). The positive pole of the power supply is tied to an external electrode, typically a mixed-metal oxide-coated titanium wire mesh. The negative pole is connected to the concrete reinforcement. The electric field drives the movement of ions in a waterborne electric current that passes between the two electrodes, resulting in the removal of the chloride ions as they are driven out of the concrete pores toward the titanium electrode. At the electrode, chlorides form chlorine gas. This strategy is effective but not perfect. One disappointment is that only about 40% of the chlorides will be extracted even after 7 weeks of treatment (Orellan, Escadeillas and Arliguie, 2004). While chloride ions are extracted from the steel reinforcement, a significant amount of alkali ions are attracted to it. This causes the pH at the reinforcement to be electrochemically passive but this can be a double-edged sword. A high alkali content may also facilitate alkali silica reactions, which could threaten the interfacial bond between the reinforcement and the concrete. In another study, it was found that a 19–33% loss in bond strength had occurred after a relatively high current density was used (Rasheeduzzafar, Ali and Al-Sulaimani,1993). While low current density is easy to apply over a large surface area, this does not prevent locally high current densities from causing damage which can manifest in the bulk material and not be limited to the reinforcement interface.

5.3. ADDING NANOPARTICLES TO THE ECE PROCESS

The idea of combining nanoparticles with ECE was first considered in 2004, when the NASA DART program sponsored a small proof of concept study that was also supported by the Corrosion Technology Group at Kennedy Space Center. Initial efforts demonstrated some positive results in terms of corrosion reductions and tensile strength increases (Cardenas *et al.* 2009). Later, an examination of corrosion in steel- reinforced cylinders demonstrated reduced corrosion rates (Cardenas and Kupwade-Patil, 2007). The study used 24-nm pozzolanic nanoparticles that were driven into the pores of relatively immature concrete and directly onto the reinforcement. The nanoparticles acted as a barrier to chloride-induced corrosion. That work also detected the first direct evidence of a conversion reaction in which calcium hydroxide content was reduced by ~8 %, indicating that C–S–H had formed in some portion of the porous network. These observations revealed a 25% increase in the tensile strength of the treated concrete. Soon after, a long-term study of the corrosion of 6.5 foot beam structures was initiated. The following sections describe how this work advanced our understanding of this application of nanomaterials to concrete

The beams were made of Portland cement Type I with a water to cement (w/c) ratio of 0.5. The dimensions were, length: 78 inches; width: 6.6 inches; height: 7.0 inches. For reinforcement, each beam had two No. 4 (0.5 inch diameter) reinforcing bars and two No.6 (0.7 inch diameter) bars in the longitudinal direction. The No. 4 bars were set to act in compression. The No. 6 bars were in tension. Shear stirrups 0.4 inches in diameter were used at 4.0 inch intervals throughout the rebar cage. Sodium chloride was dosed into the mix to simulate the use of beach sand. Dosage was set to achieve a 3.5 wt% pore content in the pore fluid, assuming it occupied 15% capillary porosity.

The beams were subjected to initial saltwater exposure dosed with 4.7 wt% NaCl. The circuit used for all treatments is illustrated in Figure 5.1. As shown in Figure 5.1 each beam was sandwiched in an assembly of three layers: 1/2-inch cellulose sponge; mixed metal oxide coated titanium mesh; and 1/2-inch plexi-glass. This 3-layer wrap was placed only on the bottom and sides of each beam. The treatments used were

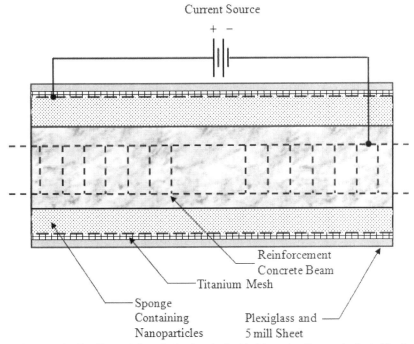

FIGURE 5.1. Circuit used for electrochemical chloride extraction and electrokinetic nanoparticle treatments. Image reproduced from Kupwade-Patil, K. (2010). Mitigation of chloride and sulfate based corrosion in reinforced concrete via electrokinetic nanoparticle treatment, Ph.D. Thesis, Louisiana Tech University, Ruston, LA.

conducted in three different dosages of 0.16, 0.33 and 0.65 liters of nanoparticle delivered per square meter of beam surface area, as measured along the bottom surface only. Variation in dosage was conducted to examine treatment efficiency and its impact on corrosion protection. A fluid suspension carrying nanoparticles was pumped to the sponge between the Plexiglas and the beam at the high end. The fluid run off was collected at the low end and re-circulated. Electrokinetic transport of particles into the beam was carried out at a constant current density of $1 A/m^2$. The positive pole of the power supply was connected to the titanium mesh, while the negative pole was connected to the concrete reinforcement. The positively-charged particles were stripped out of the fluid flowing through the sponge and driven into the concrete pores. Initially, ECE treatment was conducted for 2 weeks. For an additional 6 weeks, an electrokinetic nanoparticle (EN) treatment was conducted. Companion specimens were subjected to a full 8 weeks of ECE treatment only. Following the electrokinetic treatments, all specimens were exposed to 24 weeks of post-treatment simulated seawater, in which they were immersed up to half the height of the beam. Electrochemical corrosion potentials and corrosion rate measurements using linear polarization resistance (LPR) were taken periodically.

Corrosion current densities were measured via Linear Polarization Resistance (LPR) method and used to assess corrosion activity (Jones, p. 156, 1992). These measurements were conducted using a Solartron potentiostat (model no. 1287, manufactured by Roxboro Group Company, UK). Each measurement was conducted at a scan rate of 0.1 mv/s and ranged ±20 mV. Polarization resistance (R_p) is the ratio of change in potential (ΔE) to the change in current density (Δi) as shown below.

$$R_p = \frac{\Delta E}{\Delta i} \tag{5.1}$$

The corrosion current density (I_{corr}) can be calculated using the R_p and the constant B as follows.

$$I_{corr} = \frac{B}{R_p} \tag{5.2}$$

I_{corr} can be used to calculate the corrosion rate (CR) which is based on Faraday's law and is given by (ASTM G102-89):

$$CR = \frac{K_1 \times I_{corr} \times EW}{d} \tag{5.3}$$

where CR is the corrosion rate in mils per year, K_1 is Faraday's constant as used for corrosion rates, EW is the equivalent weight, and d is the metal density. Mercury intrusion porosimetry (MIP) was used to investigate the pore structure. These tests were carried out using the Micromertics Autopore IV 9500 in high-pressure mode, which delivered a peak value of 226 MPa. The acid-soluble chloride contents were measured in accordance with ASTM C1152. Additional characterization was performed using X-Ray diffraction, Fourier Transform Infrared Spectroscopy (FTIR) and Raman Spectroscopy.

5.4. NANOPARTICLE PACKING AND CORROSION PERFORMANCE

As mentioned earlier, the corrosion rate of embedded reinforcement can be measured in terms of corrosion current density passing through the surface of the metal. The corrosion current density (I_{corr}) was measured at 2, 4, and 11 months following the start of post-treatment saltwater immersion. As shown in Figure 5.2, I_{corr} values of 0.014, 0.011 and 0.019 µA/cm^2 were observed following 11 months for the dosages of 0.16, 0.33 and 0.65 l/m^2, respectively. The ECE and control specimens exhibited 0.053 and 2.12 µA/cm^2 at the end of 11 months. An average of 0.014 µA/cm^2 was observed among the EN-treated specimens as compared to the untreated controls, which exhibited a value of 2.12 µA/cm^2. The corrosion current density of the untreated controls was higher by a factor of 150 as compared to the EN-treated beams. A guideline for evaluating the relative severity of these corrosion current densities is provided in Table 5.1. The untreated control beams exhibited a moderate to high corrosion severity. The EN-treated and ECE-treated beams were passive. When compared more closely, it was observed that the

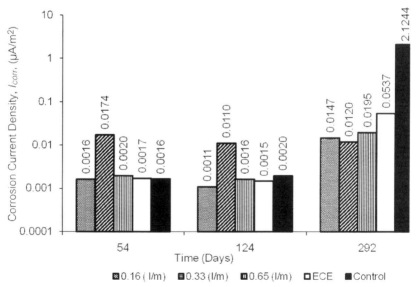

FIGURE 5.2. Corrosion current density (I_{corr}) in micro amperes per square meter for EN-treated, ECE-treated, and untreated control beams. Image adapted from Kupwade-Patil, K. (2010). Mitigation of chloride and sulfate based corrosion in reinforced concrete via electrokinetic nanoparticle treatment, Ph.D. Thesis, Louisiana Tech University, Ruston, LA.

TABLE 5.1. Severity of Corrosion Current Density.

Corrosion Current Density ($\mu A/cm^2$)	Severity
< 0.1	Passive
0.1–0.5	Low to moderate
0.5–1	Moderate to high
> 1	High

EN-treated beams exhibited a corrosion current density that was a factor of 3.5 lower than the ECE-treated beams.

5.5. NANOPARTICLES AND THE CHLORIDE BARRIER

Figure 5.3 shows the chloride content observed at various distances from the reinforcement. At distances greater than 1.5 inches from the reinforcement, the control beams exhibited chloride content that was relatively low and close in value to the treated beams. Closer to the reinforcement, the chloride content was significantly higher for these beams and were well above the American Concrete Institute (ACI) 222 code limit for new construction (ACI, 2011). All the other cases showed values that were below the ACI limit, although the scatter in the data appeared to increase at locations closer to the reinforcement. On average, the control beams exhibited higher chloride content at all locations by a factor of 4 compared to EN-treated cases. Much like the control beams, the ECE-treated beams also started to separate from the treated cases at distances within 1.5 inches from the reinforcement.

The best comparison of corrosion activity to chloride content is at the 11 month point (292 days) in Figure 5.2. The corrosion rate measurements at 11 months relate to those chloride contents in Figure 5.3 that are within 0.5 inches of the reinforcement. This content was determined at the 11 month point. When considering the columns in Figure 5.2 that represent 11 months, and those in Figure 5.3 that represent distances within 0.5 inches of the reinforcement, the rankings of treatment outcomes are the same.

Interestingly, the highest dosage, 0.65 l/m^2, did not provide very consistent or efficient protection as compared with the two lower dosages. Similarly, the high dose did not exhibit an apparent advantage in terms of the corrosion current densities shown in Figure 5.2. The observations suggest that providing a particle barrier of greater than 1 inch around the reinforcement may not be cost-effective.

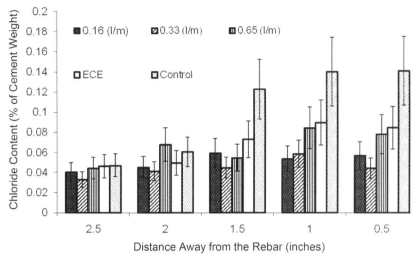

FIGURE 5.3. Chloride contents at selected distances from reinforcement. Image adapted from Kupwade-Patil, K. (2010). Mitigation of chloride and sulfate based corrosion in reinforced concrete via electrokinetic nanoparticle treatment, Ph.D. Thesis, Louisiana Tech University, Ruston, LA.

When correlated together after 11 months of exposure the controls exhibited a chloride content above the threshold limit and I_{corr} values greater than 1 µA/cm^2, both clearly indicating a high-corrosion condition. In contrast, near the reinforcement the EN-treated specimens showed lower chloride contents (below the ACI limit), and lower I_{corr} values, indicating passivity. These observations show that the nanoparticles inhibited the re-entry of the chlorides during the post-treatment exposure period.

5.6. PORE VOLUME AND STRUCTURE REVISIONS

To characterize the nature of the pore structures in the beam, MIP analysis was conducted. The most basic parameter obtained from these curves was bulk porosity. The largest pore diameter at which significant intruded mercury volume is first detected is known as the threshold diameter. This pore size gives initial significant access to the interior of the specimen. Another common parameter from MIP is the critical pore diameter, through which maximum general access to virtually the entire pore system is available.

The cumulative intruded volume versus pore diameter for the EN-treated, ECE-treated and control cases are plotted in Figures 5.4 and 5.5. Three parameters from these curves were examined: the threshold pore size, the critical pore diameter, and the bulk porosity. They were obtained from the curves and are listed together in Table 5.2. An additional bulk porosity determination method by simple water loss is also listed. The EN-treated beams showed a factor of 8 reduction in average threshold diameter (198 μm) as compared to the controls (1725 μm), and a factor of 2 below that of the ECE beams (401 μm). The particle dosage of 0.16 (l/m^2) exhibited the smallest threshold pore diameter, only 119 μm at the rebar/concrete interface. Two inches away, the 0.65 (l/m^2) dosage exhibited the smallest value of 64 μm. At the reinforcement, the treated specimens provided a critical pore diameter reduction that was on average a factor of 4 lower than the controls and 2.5 times lower than that observed among the ECE beams. Here also, the EN-treated beams showed a remarkable 74% porosity reduction. The highest dosage also posted an even more remarkable porosity reduction of 82% two inches away.

Over a long-term treatment several ionic species can be driven toward the reinforcement along with nanoparticles. In the work discussed here, the ionic species consisted of various oxidation states of sodium,

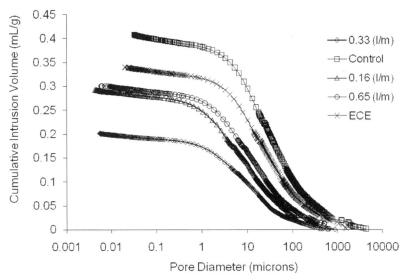

FIGURE 5.4. Mercury intrusion porosimetry (MIP) curves on powdered samples harvested adjacent to the reinforcement. Image adapted from Kupwade-Patil, K. (2010). Mitigation of chloride and sulfate based corrosion in reinforced concrete via electrokinetic nanoparticle treatment, Ph.D. Thesis, Louisiana Tech University, Ruston, LA.

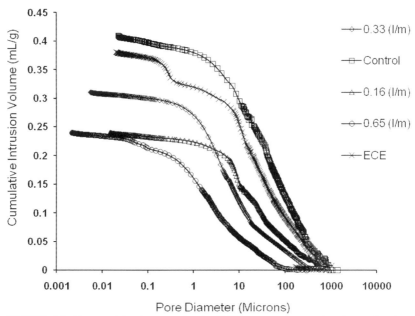

FIGURE 5.5. Mercury intrusion porosimetry (MIP) curves on powdered samples harvested two inches away from the reinforcement. Image adapted from Kupwade-Patil, K. (2010). Mitigation of chloride and sulfate based corrosion in reinforced concrete via electrokinetic nanoparticle treatment, Ph.D. Thesis, Louisiana Tech University, Ruston, LA.

potassium and calcium ions. The result shows that the combination of nanoparticles together with ionic delivery provided by ECE greatly enhanced the pore packing efficiency.

The reduction in threshold pore diameters of treated specimens demonstrated that nanoparticles effectively reduced access to the material as whole by forming a physical barrier that inhibited diffusion of chlorides. The chloride contents and porosities observed indicated that the nanoparticle treatments held the chloride content below the ACI threshold level for new construction by removing the chlorides from the pores and modifying the pore structure to inhibit their re-entry. Microstructural characterization in the next section describes the phases formed due to the EN treatment.

5.7. MICROSTRUCTURAL PHASES GENERATED BY NANOPARTICLES

We can start by examining corrosion phases that were suppressed by the nanoparticle treatments. Fracture surfaces presented in Figure 5.6

TABLE 5.2. Porosity and Pore Structure in Nanoparticle-Treated Beams.

Treatment Type	MIP				Porosity			
	Threshold Pore Diameter (μm)		Critical Pore Diameter (μm)		MIP (%)		WLR (%)	
	R/C Surface	2″ from R/C	R/C Surface	2″ from R/C	R/C Surface	2″ from R/C	R/C Surface	2″ from R/C
Control	1725	773	29	18	11.4	15.3	8.9	12.9
ECE	401	482	19	12	9.4	10.4	13.8	9.1
0.16 (l/m)	119	352	6	11	7.2	8.4	5.2	4.9
0.33 (l/m)	292	294	7	6	4.1	4.9	4.8	5.1
0.65 (l/m)	184	64	10	2	3.01	2.8	5.3	6.4

R/C : Rebar concrete interface

show the imprints of the rebar/concrete interface. The control sample exhibited significant amounts of reddish-colored corrosion products on the imprint. This reddish color was not evident on the EN-treated cases. Some of the analyses conducted in this section identify the chemical phases of these corrosion products. Iron-oxides have a detrimental effect on a given structure because their formation is accompanied by volume expansions that induce tensile stress in the concrete, which in turn leads to structural failure.

Raman spectrography was conducted on beam samples to examine the phases present. Raman spectra were run on EN- and ECE-treated beam samples, as well as the control beam samples. These spectra are shown in Figure 5.7. For the EN-treated specimens the analysis revealed sharp peaks that indicated C–S–H in the range of 400–600 cm^{-1}. By comparison, the presence of weak C–S–H peaks on the control and the ECE beams was an indication of normal levels of occurrence for these phases, stemming from the initial hydration reactions that followed mixing with water (Mindess, Young and Darwin, 2002).

FIGURE 5.6. Imprints of corrosion products at the rebar/concrete interface for EN-treated and control specimens (Kupwade-Patil, et al. 2008). Image reproduced from Kupwade-Patil, K. (2010). Mitigation of chloride and sulfate based corrosion in reinforced concrete via electrokinetic nanoparticle treatment, Ph.D. Thesis, Louisiana Tech University, Ruston, LA.

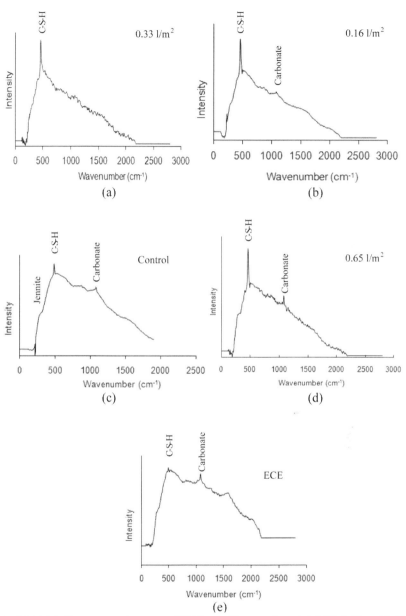

FIGURE 5.7. Raman spectroscopy of EN-treated beams dosed at 0.65, 0.33 and 0.16 l/m²). Spectra for ECE-treated and untreated control beams are also shown. Image reproduced from Kupwade-Patil, K. (2010). Mitigation of chloride and sulfate based corrosion in reinforced concrete via electrokinetic nanoparticle treatment, Ph.D. Thesis, Louisiana Tech University, Ruston, LA.

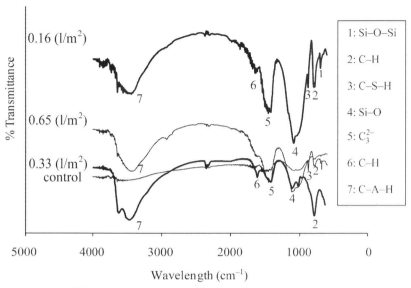

FIGURE 5.8. FTIR transmission spectra EN-treated specimens and untreated control. Image adapted from Kupwade-Patil, K. (2010). Mitigation of chloride and sulfate based corrosion in reinforced concrete via electrokinetic nanoparticle treatment, Ph.D. Thesis, Louisiana Tech University, Ruston, LA.

The sharp peaks of the EN-treated cases indicated the formation of additional C–S–H through the conversion of calcium hydroxide occurring in a typical pozzolanic reaction. The additional C–S–H formation was probably due to the reaction between the silica component of the nanoparticles and the $Ca(OH)_2$ that is abundantly available within the hydrated state of Type I Portland cement.

Fourier Transform Infrared (FTIR) transmittance spectra from ground samples of beams are plotted in Figure 5.8. The EN-treated specimens showed broad peaks in the $3400–3600$ cm^{-1} range. These bands stem from OH bond vibrations of water molecules found in mono-calcium aluminate hydrate, di-calcium aluminate hydrate and tri-calcium aluminate hexahydrate (Carrasco, Ruiz and Miravitlles, 2008). In general, these phases are collectively referred to as C–A–H. More complex spectras of calcium silicate hydrate were found at $900–1100$ cm^{-1} (Yu et al., 1999). The EN-treated cases exhibited Si–O–Si bond vibrations associated with C–S–H in the $600–700$ cm^{-1} range, as well as Si–O bonds also indicative of C–S–H in the 900-1100 cm-1range. Carbonate vibration peaks were also manifest at 1440 cm^{-1}.

X-ray diffraction analyses were conducted to examine the presence of crystalline phases. A collection of spectra taken from beam samples are presented in Figure 5.9. For the EN treated specimens small peaks of monocalcium aluminate (CA) were located at 2θ values of $18°$. No such peaks were observed in the ECE-treated cases or the control beams. All beams exhibited peaks indicating portlandite, quartz and calcium silicate hydrate (C–S–H) at 2θ values of $20°$, $27°$ and $30°$.

Corrosion products of iron were also detected via XRD. These were observed among the control and ECE beams, which exhibited lepidocrocite, an iron-corrosion product. Goethite, another product of iron corrosion, was only observed in the control beams. This indicated that the post-saltwater exposure led to corrosion that was more advanced than observed within the ECE-treated beams (which exhibited only lepidocrocite). The EN-treated beams did not show any signs of corrosion product phases. The Raman, FTIR and XRD analyses also indicated that EN-treatments were forming additional C–S–H and C–A–H phases, in keeping with pozzolanic reactivity expectations.

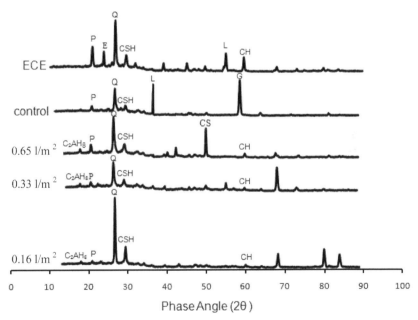

C_3AH_6: Tricalcium Aluminate hexahydrate (CA), P: Portlandite, E: Ettringite, Q: Quartz, G: Goethite, L: Lepidocrocite, CS: Calcium Silicates (C_2S and C_3S)

FIGURE 5.9. XRD analysis of powdered beam samples. Image adapted from Kupwade-Patil, K. (2010). Mitigation of chloride and sulfate based corrosion in reinforced concrete via electrokinetic nanoparticle treatment, Ph.D. Thesis, Louisiana Tech University, Ruston, LA.

5.8. SUMMARY OF NANOPARTICLE ASSISTED CORROSION MITIGATION

EN-treatment exhibited applicability to large structures. It utilized a weak electric field to transport 24-nm alumina-coated silica nanoparticles directly to the steel reinforcement. These nanoparticles appeared to form an effective physical barrier against chloride re-penetration. The treatments were successful in mitigating the moderate to high reinforcement corrosion observed in the untreated beams. Untreated controls exhibited corrosion rates that were 150 times higher than the EN-treated specimens. At the rebar interface, EN-treated beams also exhibited chloride content that was 4 times lower. Nanoparticles reacted with available calcium hydroxide to form additional C–S–H and C–A–H. As EN dosages were increased, porosity reduction improved. In general, EN treatments exhibited 2.5–3 times the porosity reduction obtained by ECE treatments. ECE treatment was found to be less effective than EN treatment in preventing the return of chlorides. At the 11-month mark, the corrosion rate within ECE- treated beams was 3.5 times the corrosion rate of the EN-treated beams.

5.9. REFERENCES

ACI 222.3-R11, *Guide to Design and Construction Practices to Mitigate Corrosion of Reinforcement in Concrete Structures,* American Concrete Institute, Farmington Hills, MI, USA, April 2011.

Broomfield, J. (1997) *Corrosion of Steel in Concrete: Understanding, Investigation and Repair,* 2nd Edition, E & FN Spon Publications, UK, pp. 16–28.

Bentur, A., Diamond, S. and Berke, N. (1998) *Steel Corrosion in Concrete,* 1st Edition, E & FN Spon Publications, UK, pp. 24–38.

Cardenas, H.E., Alexander, J.B., Kupwade-Patil, K., Calle, L.M., Field Testing of Rapid Electrokinetic Nanoparticle Treatment for Corrosion Control of Steel in Concrete, National Aeronautics and Space Administration, Technical Report, NASA/TM-2009-214761, August 2009.

Cardenas, H. and Kupwade-Patil, K. (2007) "Corrosion Remediation Using Chloride Extraction Concurrent with Electrokinetic Pozzolan Deposition in Concrete," *New Solutions for Environmental Pollution, Sixth Symposium on Electrokinetic Remediation,* Vigo, Spain, pp. 117.

Carrasco L., Ruiz, J. and Miravitlles, C. (2008) "Supercritical Carbonation of Calcium Aluminate Cement," *Cem. and Concr. Res.,* Vol. 38, pp. 1033–1037.

Goli, N., "Investigation of Nanoparticle Technology for Corrosion Mitigation in Reinforced Concrete," Masters Thesis, Louisiana Tech University, Ruston, LA, May 2005.

Jones, D., Principles and Prevention of Corrosion, Macmillan Publishing Co., New York, 1992.

Koch, G. K. *et al.* (2002) "Corrosion Cost and Preventive Strategies in the United States," US Department of Transportation Federal Highway Administration, Report No FHWA-RD-01-156, March.

Kupwade-Patil, K. *et al.* (2008) "Corrosion Mitigation in Concrete Using Electrokinetic Nanoparticle Treatment," *Proceeding of Excellence in Concrete Construction through Innovation,* London, UK, pp. 365–371.

Marcotte, T., Hansson, C. and Hope, B. (1999) "The Effect of Electrochemical Chloride Extraction Treatment on Steel-Reinforced Mortar, Part II: Microstructural Characterization," *Cement and Concrete Research,* Vol 29, pp. 1561–1568.

Mindess, S., Young, F. and Darwin, D. (2002). *Concrete,* Prentice Hall, 2nd Ed, New Jersey.

Orellan, J., Escadeillas, G. and Arliguie G. (2004) "Electrochemical Chloride Extraction: Efficiency and Side Effects," *Cement and Concrete Research,* Vol 34, pp.227–234.

Rasheeduzzafar, A., Ali, G. and Al-Sulaimani, G. (1993) "Degradation of Bond Between Reinforcing Steel and Concrete due to Cathodic Protection Current," *ACI Materials Journal,* Vol. 90, pp. 8–15.

Yu, P. *et al.* (1999) "Structure of Calcium Silicate Hydrate (C–S–H) Near-, Mid-, and Far-Infrared Spectroscopy," *Journal of American Ceramic Society,* Vol 82. No. 3, pp. 742–748.

CHAPTER 6

Sulfate Removal and Damage Recovery

ULFATE ATTACK, a profound form of degradation, is considered
preventable but expensive to repair. The remedy usually involves
removal and replacement of structures. Degradation begins with the dif-
fusive invasion through concrete pores of sodium, calcium or magnesium sulfate from soils or ground water (Skalny, Marchand and Odler,
2001; Shazali, Baluch and Al-Gadhib, 2001; Neville, 1996). Damage
occurs in the form of volume expansion, leaching, spalling, delamina-
tion, cracking, and increased permeability. Expansion is driven by the
formation of ettringite and gypsum hydrate (calcium sulfate) (Mehta,
1983; Collepardi, 2001; Bing and Cohen, 2000; Rasheeduzzafar *et al.*,
1994; Neville, 2004). The gypsum is produced when sulfate ions react
with calcium hydroxide (C–H), which is naturally abundant in ordinary
Portland cement. Ettringite formation involves sulfates reacting with
the monosulfates resident in concrete. Ettringite takes the form of 7–9
µm long hexagonal needles that push the cement binder apart as they
grow. This causes tensile stress, volume expansion, and serious cracking. The following sections describe how sulfate attack is currently
handled, and introduces a remarkable paradigm shift in the concept of
recovery. The application of nanomaterials in this area has created an
exciting opportunity that, up to now, has not been considered.

6.1. SULFATE ATTACK PREVENTION

In current practice, Type V Portland cement (which is sulfate-resis-
tant) and admixtures of pozzolans with Type I or II Portland cements,
are used to avoid sulfate attack (Ghafoori and Mathis, 1997; Monterio

and Kurtis, 2003; Naik *et al.*, 2006). Type V cement reduces the amount of tri-calcium aluminate (C_3A), the key reactant in ettringite generation, present in the mixture. This is the chief line of defense. Another method is porosity reduction, achieved mainly by mixing with low water/cement (*w/c*) ratios, so that sulfate diffusion is inhibited. For *w/c* ratios above 0.45, Type V cements fend off sulfate attack better than Type I cements, because the sulfate susceptibility of Type V is generally not sensitive to w/c ratio. This is not true of Type I and Type II cements (Cohen and Bentur, 1988; Cohen and Mather, 1991).

Pozzolans are also used to prevent sulfate attack. In particular, fly ash and silica fume admixtures are used to enhance resistance (Tikalsky and Carrasquillo, 1992). As noted before, these admixtures react with calcium hydroxide, causing it to be unavailable for reaction with sulfates. Pozzolanic reactivity also reduces volume porosity, thus discouraging the diffusion of sulfates in the porous volume.

Until recently, no remedy except replacement was available for damage already done. The possibility of simply rolling back the impact of sulfate attack was unknown. But application of nanomaterials and electrochemical methods has achieved non-destructive recovery from sulfate attack damage. The following sections describe how nanoparticle treatment was used to stop this chemical attack and reverse some of its damaging effects.

6.2. USING NANOPARTICLES TO ADDRESS SULFATE ATTACK

The new repair strategy applies electrochemical sulfate extraction simultaneously with the injection of positively-charged nanoscale pozzolans. This combined approach is similar to concepts used in treating corrosion of reinforcements, except that here the treatment is applied to bulk material, where the degradation occurs. The first attempt at this new approach was conducted in 2007, using the same 24-nm alumina-coated silica particle that was used in the treatment of reinforcement corrosion (Cardenas *et al.* 2011). In the present case, the nanoparticle injection was intended not only to reduce porosity for the prevention of sulfate re-entry, but also to strengthen the damaged concrete.

Cylindrical concrete specimens 3 × 6 inches in size were fabricated, in accordance with ACI 211.1, using a mix with a 3/8-inch maximum aggregate size (Table 6.1) (ACI, 1991). The mix was prepared in accordance with ASTM C192 using a w/c ratio of 0.5 (ASTM, 2001). Electrodes of mixed-metal oxide-coated titanium wire 5 inches in length were cast down the center of each specimen. Limewater curing was conducted for 7 days. For concrete specimen immersion, a 35.5 g/l solution of anhydrous sodium sulfate was used. Solution/specimen volume ratio of 4/1 was maintained for the immersion of the specimens at 23°C as per ASTM C 1012 (ASTM, 2001). This exposure solution was changed out weekly.

The period of sodium sulfate exposure was 30 days. Batch companions were kept in lime water to serve as undamaged control specimens with no sulfate exposure. After 30 days, the electrokinetic nanoparticle treatments began to extract the negatively-charged sulfate ions while injecting positively-charged nanoparticles. The positive pole of the power supply was connected to the exterior electrode and the negative pole was connected to the embedded wire electrode. A current density calculated at 1 A/m^2 over the outside surface area of the cylinder was maintained during the treatment. After 7 days of sulfate extraction, EN treatment was applied for 7 more.

Compression testing was done as per ASTM C 684-99 (ASTM, 2001). Prior to strength testing, the inner electrode was trimmed flush

TABLE 6.1. Sulfate Attack Specimen Design.

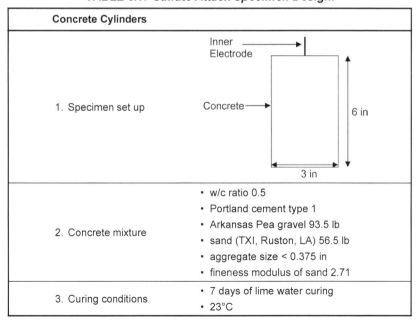

Concrete Cylinders	
1. Specimen set up	Inner Electrode, Concrete, 6 in, 3 in
2. Concrete mixture	• w/c ratio 0.5 • Portland cement type 1 • Arkansas Pea gravel 93.5 lb • sand (TXI, Ruston, LA) 56.5 lb • aggregate size < 0.375 in • fineness modulus of sand 2.71
3. Curing conditions	• 7 days of lime water curing • 23°C

Adapted from Cardenas et al. 2011 with permission from ASCE.

with the top of the specimen. All specimens regardless of treatment history were batched with the same inner-electrode. Samples of strength-tested specimens were subjected to MIP, SEM, EDAX, Raman spectroscopy and FTIR analysis.

6.3. ASSESSING SULFATE DAMAGE

Figure 6.1 shows a cylinder specimen after 28 days of exposure to sodium sulfate. Areas of gross cracking and spallation are boxed. The dimensions of the sulfate-exposed specimens were monitored during the exposure period.

Expansion in terms of change in length is plotted over time in Figure 6.2. After the first week of exposure, expansion occurred at the relatively steady rate of about 0.5% per week. The progress of sulfate attack on compressive strength in both exposed and unexposed cylinders were obtained weekly and are listed in Figure 6.3. Differences in strength appeared after the second week of exposure—which correlated to a similar observation in length-change shown in Figure 6.2. This delay is not too surprising, because diffusion and crystal growth take time to cause damage. The differences in strength increased through weeks 3 and 4 of exposure. Strength decreased by 13% at the end of week 3, and a

Spallation Damage

FIGURE 6.1. Spallation damage due to sulfate exposure. Adapted from Cardenas et al. 2011 with permission from ASCE.

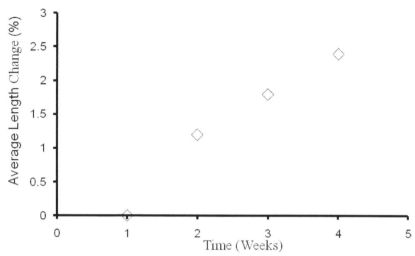

FIGURE 6.2. Length changes due to sulfate-induced expansion. Reproduced from Cardenas et al. 2011 with permission from ASCE.

total of 38% by the end of week 4. These damage outcomes were in line with other work, in which a 40% decrease in compressive strength was observed after 50 days (Boyd and Mindess, 2004). It was further shown that ettringite crystals were responsible for crack formation (Nielsen, 1966).

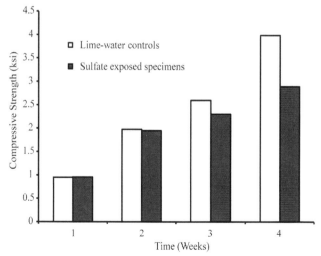

FIGURE 6.3. Influence of sodium sulfate exposure on compressive strength. Adapted from Cardenas et al. 2011 with permission from ASCE.

6.4. TREATMENT IMPACT ON SULFATE ATTACK

While nanoparticle treatment was ongoing, companion specimens were removed from sulfate exposure and placed in limewater. Following 2 weeks of treatment, all these controls, as well as the treated specimens, were subjected to compressive strength testing. On average, the EN-treated specimens exhibited 33% higher compressive strength than those exposed to sulfate. See Table 6.2.

Table 6.2 contains pore structure data obtained from MIP analysis. Table 6.3 provides the pore diameters generally expected for construction ceramics of various types. Figure 6.4 contains the pore-size distribution curves obtained for the test cases of concrete subjected to sulfate attack and recovery treatment, as well as those used for control comparison. Threshold pore diameters obtained from Figure 6.4 are listed in Table 6.2. The sulfate-exposed sample had threshold pore sizes of 221 μm—the high end of the typical range for concrete (Winslow 1989). By comparison, the treated specimens exhibited threshold pore size values of 58 μm, smaller by over a factor of 5. This new threshold pore size also compared favorably to the undamaged control specimens in limewater, being lower by a factor of 2.

As the reader may recall from the section on corrosion of reinforcements, pozzolanic nanoparticles reduce the available $Ca(OH)_2$ by converting it to C–S–H or C–A–H. This makes it less available for damaging reactions with sulfates. The reaction, however, is limited by the fact that the sulfates reach the calcium hydroxide before the nanoparticles arrive. Fortunately, even without these beneficial pozzolanic reactions, the simple loading of pores with nanoscale pozzolans can reduce pore volume. It is well-recognized that smaller pore volume tends to increase strength and inhibit diffusion of ionic species, such as sulfates. This is true even when the new material is somewhat weaker than the surrounding solid phases (Neville 1996). Table 6.2 shows that, on av-

TABLE 6.2. Compressive Strength and Pore Structure.

Type of Exposure	Compressive Strength (psi)	Porosity by MIP (%)	Threshold Pore Diameter (μm)
Lime water	4000	31	123
Sulfate-exposed	2100	40	221
EN-treated	2800	29	58

Adapted from Cardenas *et al.* 2011 with permission from ASCE.

TABLE 6.3. Typical Pore Diameter Ranges for Concrete Materials.

Material Type	Diameter (µm)
Cement paste	0.001–1.4
Concrete	0.001–250
Aggregate	0.01–100
Concrete block	0.01–750

Reproduced from Cardenas *et al.* 2011 with permission from ASCE.

erage, Na_2SO_4-exposed specimens exhibited a 39% volume porosity and a compressive strength of only 2100 psi. By contrast, EN-treated specimens exhibited a 26% volume porosity and a higher strength of 2800 psi. The EN-treated specimens thus exhibited a 33% higher strength and a 13% lower porosity than the sulfate-exposed cylinders. The nanoparticle treatments were successful in increasing the strength of sulfate-damaged concrete by reducing its porosity.

The threshold pore size of the concrete was also examined. Limewater control specimens had a threshold pore size of 123 µm and a porosity of 31%. Specimens exposed to sulfates developed cracks, leading to a threshold pore size increase from 123 µm to 221 µm. In addition, porosity increased to 40%, which is up from the 31% porosity observed among the limewater controls. Following nanoparticle treatment, the porosity dropped to 29% and the threshold pore size fell even more dramatically down to 58. These observations demonstrate that after sul-

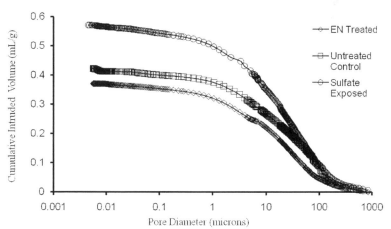

FIGURE 6.4. *Volume porosity and pore-size distributions obtained using mercury intrusion porosimetry (MIP). Reproduced from Cardenas et al. 2011 with permission from ASCE.*

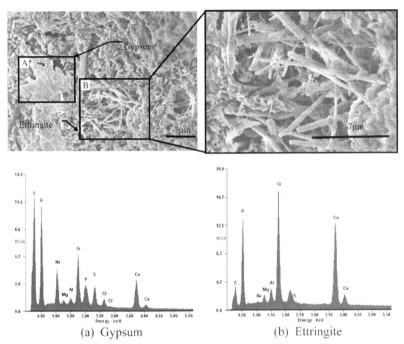

FIGURE 6.5. Ettringite formation in sulfate exposed concrete. Reproduced from Cardenas et al. 2011 with permission from ASCE.

fate attack expanded both threshold pore size and volume porosity, the nanoparticle treatment dramatically reduced both values to levels that were actually below those of the undamaged specimens.

The SEM images in Figure 6.5 show two views of the long prismatic needles of ettringite crystals. These images were taken from a sample removed from sodium sulfate-exposed specimens and EN-treated specimens. Figure 6.5(a) and 6.5(b) (higher magnified image) exhibit the formation of needle-like ettringite crystals (labeled B) and gypsum (labeled A). These phases were observed abundantly within the sulfate-exposed specimens. Significant void space is also evident, indicating the locations where porosity increases were induced by expanding ettringite needles. The EDAX spectra were run on these areas of gypsum and ettringite (these are the boxed regions labeled A and B). The scans detected the presence of aluminum, calcium and sulfur, which correlate to both of these phases (Mindess, Young and Darwin, 2002; Sahmaran, Erdem and Yaman, 2006). Expansion and damage caused by gypsum and ettringite formation accounts for the increases in porosity and threshold pore diameters listed in Table 6.2. Further evidence of this damage was observed in the strength reductions also listed in Table 6.2. Based on these observations, sulfate exposure caused the concrete to exhibit classic signs of sulfate attack.

Figure 6.6 shows the EDAX analyses run on polished samples taken from the cylinders. It was found that the sulfate-exposed specimens contained a 2.9 weight % content of sulfur and a 4.8% content of sodium. By comparison, the treated specimens contained only trace amounts, less than 1%, of both of these species, which indicates that the electro-kinetic treatments had removed significant amounts of them. In general, it would not be expected that the sodium content should be so low in a given concrete specimen. It is probable that the low sodium content was due to the fact that the electric field tends to draw positive ions, such as sodium, to the central titanium electrode. Because the electrode was further into the specimen than the locations examined via BSE, the amount of sulfates appeared to be reduced. The treatment was effective in reducing the content of the sulfates as well as other ions.

Raman spectra representing the condition of each specimen are shown in Figure 6.7. The sulfate-exposed case showed a peak for S–O bond vibration at 1083 cm^{-1}, which is associated with sulfate (Kirkpatrick *et al.*,1997). In addition, a minor peak associated with C–S–H was noted at 455 cm^{-1}. This peak generally lies in the broad bandwidth of 400–600 cm^{-1}. In the control specimen spectra of Figure 6.7, an apparent thaumasite peak was observed at 1072 cm^{-1}. This peak is viewed as a combination of carbonate and sulfate group signatures (Potgieter-Vermaak, Potgieter and Van Grieken, 2006; Brough and Atkinson, 2001). As shown also in Figure 6.7, the EN-treated specimens showed sharp C–S–H peaks at 459 cm^{-1}.

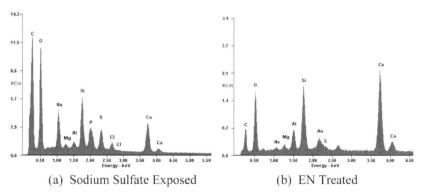

(a) Sodium Sulfate Exposed   (b) EN Treated

FIGURE 6.6. BSE image and EDAX polished sulfate exposed and EN treated specimen. Adapted from Cardenas et al. 2011 with permission from ASCE.

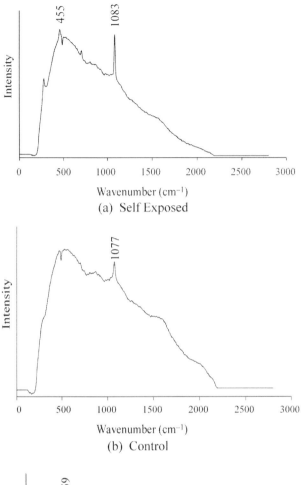

(a) Self Exposed

(b) Control

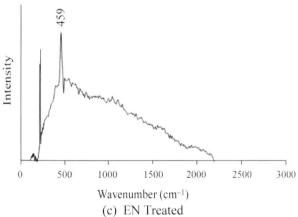

(c) EN Treated

FIGURE 6.7. Raman spectroscopy of EN-treated and control specimens. Adapted from Cardenas et al. 2011 with permission from ASCE.

The FTIR spectra of both sulfate-exposed and EN-treated samples are plotted in Figure 6.8. For the treated specimens, broad band peaks in the range of 3400–3600 cm^{-1} indicated the OH vibrations of water molecules in monocalcium aluminate hydrate, di-calcium aluminate hydrate, and tri-calcium aluminate hexahydrate (Carrasco, Ruiz and Miravitlles, 2008). The presence of C–H at 1527 cm^{-1}.was also detected in these samples. Among the untreated samples appeared C–S–H as expressed by Si–O–Si and Si–O vibration peaks at the 698 cm^{-1} and 1120 cm^{-1} locations. Also, strong ettringite peaks were observed in sulfate-exposed specimens, as well as in the untreated controls at the 877 cm^{-1} location (Pajares, Ramirez and Varela, 2003). By comparison, the EN-treated samples exhibited relatively weak ettringite peaks. Peaks of CO_3^{2-} at 1440 cm^{-1} signaled the presence of carbonate in the untreated controls (Carrasco, Ruiz and Miravitlles, 2008). In addition all the samples exhibited SO_4^{2-} bond vibrations at 680 cm^{-1} (Carrasco, Ruiz and Miravitlles, 2008; Pajares, Ramirez and Varela, 2003)

As shown in Figure 6.8, FTIR transmittance analysis exhibited significantly weaker peaks for ettringite in the EN-treated cases as compared with the untreated controls, and for the sulfate-exposed samples. The Raman spectra in Figure 6.7 showed no such sulfate peaks for the EN-treated specimens as were present for the sulfate-exposed samples. It is likely that this reduction of sulfate content by electrokinetic extraction would have reduced the extent of monosulfate conversion, which

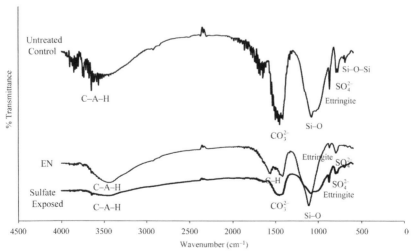

FIGURE 6.8. FTIR transmittance spectra of specimen samples. Adapted from Cardenas et al. 2011 with permission from ASCE.

would explain the relatively weak ettringite peak for the treated cases in Figure 6.8. These observations clearly suggest that the sulfates were effectively extracted, thus mitigating further ettringite formation.

The sulfate extraction process was not complete, as indicated by the presence of SO_4^{2-} bond vibrations among the treated specimens in Figure 6.8. This observation is reminiscent of similar removal limitations generally associated with chloride extraction processes (Fajardo, Escadeillas and Arliguie 2006; Orellan, Escadeillas and Arliguie, 2004; Marcotte, Hansson and Hope, 1999). The sulfate ion is considerably larger than a chloride ion. From this standpoint, it is not surprising that a relatively limited 14-day treatment would remove all the sulfates. Even so, the current treatment resulted in a 33% rise in strength that partially offset the 50% drop caused by the sulfate attack. The EN treatments were effective at reducing the severity of attack despite very limited particle dosage and treatment duration. As other work has shown, raising the current density may possibly permit more rapid and complete sulfate extraction with less treatment time.

It is notable that strength recovery seemed limited as compared to porosity reduction. Porosity of EN-treated specimens (29%) was lower than that of sulfate-exposed specimens (40%). In contrast, the limited strength recovery was disappointing. The strength of the EN-treated specimens rose only to 2800 psi. In comparison, the limewater-exposed specimens retained a strength of 4000 psi. This shortfall may be related to the macro-spallation observed in Figure 6.1. Nanoparticle treatment is clearly incapable of filling such macro defects. At best, such a treatment may only impact the micro-cracking. Based on these observations, one could maintain that nanoparticle treatments would be most beneficial before advanced stages of sulfate attack have occurred.

6.5. SUMMARY OF SULFATE DECONTAMINATION AND RECOVERY

It is clear that nanoparticle treatment can provide at least partial recovery of strength. This work showed that EN-treatment removed significant amounts of sulfates, increased strength by 33%, reduced porosity by 28%, and dropped the threshold pore diameter by a factor of 4. This suggests that it may slow the diffusion of sulfates back into the structure over time and thus inhibit the redevelopment of a sulfate attack issue. The primary benefits of this type of treatment may be greatest when they are applied before sulfate attack has reached a stage where macro-spallation or cracking has become manifest.

6.6. REFERENCES

ACI 211.1 (1991) "Standard Practice for Selecting Proportions for Normal, Heavy Weight and Mass Concrete," American Concrete Institute, Farmington Hills, MI.

Aligazaki, K. (2005) *Pore Structure of Cement Based Materials: Testing Interpretation Requirements,* Taylor and Francis, London, UK.

ASTM (2001) "Standard Practice for Making and Curing Concrete Test Specimens in the Laboratory," C 192, West Conshohocken, PA.

ASTM (2001) "Standard Test Method for Length Change of Hydraulic-Cement Mortars Exposed to Sulfate Solution," C1012-95a, West Conshohocken, PA.

ASTM (2001) "Standard Test Method for Making, Accelerated Curing, and Testing Concrete Compression Test Specimens," C 684, West Conshohocken, PA.

Bing, T. and Cohen, M. (2000) "Does Gypsum Formation during Sulfate Attack on Concrete Lead to Sulfate Attack," *Cem. Concr. Res.,* 30(1):117–123.

Boyd, A. and Mindess, S. (2004) "The Use of Tension Testing to Investigate the Effect of W/C Ratio and Cement Type on the Resistance of Concrete to Sulfate Attack," *Cem. and Concr. Res.,* 34(3):373–377.

Brough, A. and Atkinson, A. (2001) "Micro-Raman Spectroscopy of Thaumasite," *Cem. and Concr. Res.,* 31(3):421–424.

Cardenas, H., Kupwade-Patil, K., Eklund, S., "Recovery from Sulfate Attack in Concrete via Electrokinetic Nanoparticle Treatment", *American Society of Civil Engineers—Journal of Materials in Civil Engineering,* Vol. 23, No.7, July, 2011.

Cardenas, H., and Kupwade-Patil, K. (2007) "Corrosion remediation using chloride extraction concurrent with electrokinetic pozzolan deposition in concrete." *6th Symposium on Electrokinetic Remediation,* Vigo, Spain, P11, 117.

Cardenas, H. and Struble, L. (2006). "Electrokinetic nanoparticle treatment of hardened cement paste for reduction of permeability." *J. Mater., Civ. Eng,* 18 (4), 554–560.

Carrasco, L., Ruiz, J. and Miravitlles, C. (2008) "Supercritical Carbonation of Calcium Aluminate Cement," *Cem. and Concr. Res.,* 38(8–9):1033–1037.

Cohen, M. and Bentur, A. (1988) "Durability of Portland Cement-Silica Fume Pastes in Magnesium Sulfate and Sodium Sulfate Solutions," *ACI Mater. J.,* 85(18):148–157.

Cohen, M. and Mather, B. (1991) "Sulfate Attack on Concrete: Research Needs," *ACI Mater. J.,* 88(1):62–69.

Collepardi, M. (2001) "Ettringite Formation and Sulfate Attack on Concrete," *ACI Special Publication,* 200:21–38.

Fajardo, G., Escadeillas, G. and Arliguie, G. (2006) "Electrochemical Chloride Extraction (ECE) from Steel-Reinforced Concrete Specimens Contaminated by 'Artificial' Sea-water," *Corr. Sci.,* 48(1):110–125.

Ghafoori, N. and Mathis, R. (1997) "Sulfate Resistance of Concrete Pavers," *J. Mater. Civ. Eng.,* 9(1):35–40.

Gordon, K., Kupwade-Patil, K., Lee, L., Cardenas, H., and Moral, O. (2008). "Long term durability of reinforced concrete rehabilitated via electrokinetic nanoparticle treatment." *Proc. Excellence in concrete construction through innovation,* Taylor and Francis, London, pp. 373–379.

Kirkpatrick, J. *et al.* (1997) "Raman Spectroscopy of C–S–H, Tobermorite and Jennite," *Advn. Cem. Bas. Mat.,* 5:93–99.

Kupwade-Patil, K. (2007). "A new corrosion mitigation strategy using nanoscale pozzolan deposition," Master's Thesis, Louisiana Tech University, Ruston, USA.

Kupwade-Patil, K. and Cardenas, H. (2008) "Corrosion mitigation in concrete using electrokinetic injection of reactive composite nanoparticles," *Proc. 53rd International SAMPE symposium,* Long Beach, CA.

Kupwade-Patil, K., Gordon, K., Xu, K., Moral, O., Cardenas, H., and Lee, L. (2008). "Corrosion Mitigation in concrete using electrokinetic nanoparticle treatment," *Proc. Excellence in concrete construction through innovation,* London, Taylor and Francis, UK, pp. 365–371.

Marcotte, T., Hansson, C. and Hope, B. (1999) "The Effect of the Electrochemical Chloride Extraction Treatment on Steel-Reinforced Mortar, Part I: Electrochemical Measurements," *Cem. and Concr. Res.,* 29(10):1555–1560.

Mehta, P. K. (1983) "Mechanism of Sulfate Attack on Portland Cement Concrete—Another Look," *Cem. Concr. Res.,* 13:401–406.

Mindess, S. (2001) "The Strength and Fracture of Concrete: The Role of Calcium Hydroxide," in *Material Science of Concrete: Calcium Hydroxide in Concrete,* editors Skalny, J., Gebaver, J. and Odler, I., The American Ceramic Society, Ohio, 2001, pp. 143–154.

Mindess, S., Young, F. and Darwin, D. (2002) *Concrete,* Prentice Hall, 2nd Ed, New Jersey.

Monterio, P. and Kurtis, K. (2003) "Time to Failure for Concrete Exposed to Severe Sulfate Attack," *Cem. Concr. Res.,* 33(7):987–993.

Morrison. R., and Boyd, R. (1992), *Organic Chemistry,* Prentice Hall, New Jersey.

Naik, N. et al. (2006) "Sulfate Attack Monitored by MicroCT and EDXRD: Influence of Cement Type, Water-to-Cement Ratio, and Aggregate," *Cem. and Concr. Res.,* 36(1): 144–159.

Neville, A. (1996) *Properties of Concrete,* Wiley, New York.

Neville, A. (2004) "The Confused World of Sulfate Attack," *Cem. Concr. Res.,* 34(8): 1275–1296.

Nielsen, J. (1966) "Investigation of Resistance of Cement Paste to Sulfate Attack," *Highway Research Record,* 113, (1966), pp. 114–117.

Odler, I, (2000) *Special inorganic cements,* E &FN Spon, London.

Orellan, J., Escadeillas, G. and Arliguie, G. (2004) "Electrochemical Chloride Extraction: Efficiency and Side Effects," *Cem. and Concr. Res.,* 34(2):227–234.

Pajares, I., Ramirez, S. and Varela, M. (2003) "Evolution of Ettringite in Presence of Carbonate and Silicate Ions," *Cem. and Concr. Comp.,* 25(8):861–865.

Potgieter-Vermaak, S., Potgieter, J. and Van Grieken, R. (2006) "The Application of Raman Spectrometry to Investigate and Characterize Cement, Part I: A Review," *Cem. and Concr. Res.,* 36(4):656–662.

Ramachandran, V., and Beaudoin, J. (2001) *Handbook of Analytical Techniques in Concrete Science and Technology,* William Andrew Publishing, New York.

Rasheeduzzafar, O. *et al.* (1994) "Magnesium-Sodium Sulfate Attack in Plain and Blended Cements," *J. Mater. Civ. Eng.,* 6(2):201–222.

Sahmaran, M., Erdem, T. and Yaman, I. (2006) "Sulfate Resistance of Plain and Blended Cements Exposed to Wetting-Drying and Heating-Cooling Environments," *Constr. Build. Mater.,* 21(8):1771–1778.

Shazali, M., Baluch, M and Al-Gadhib. H. (2006) "Predicting Residual Strength in Unsaturated Concrete Exposed to Sulfate Attack," *J. Mater. Civ. Eng.,* 18(3):343–354.

Skalny, J., Marchand, J. and Odler, I. (2001) *Sulfate Attack on Concrete,* Taylor and Francis, London.

Taylor H. F. W. (1997) *Cement Chemistry,* 2nd Ed, Thomas Telford, London.

Tikalsky, P. and Carrasquillo, R. (1992) "Influence of Fly Ash on the Sulfate Resistance of Concrete," *ACI Mater. J.,* 89(1):69–75.

Winslow, D. (1989) "Some Experimental Possibilities with Mercury Intrusion Porosimetry, *Proc. Vol. 137: Pore Structure and Permeability of Cementitious Materials,* Materials Research Society, Warrendale, PA, pp. 93–103.

Yu, P., Kirkpatrick, J., Poe, B., McMillan, P., and Cong, X. (1999) "Structure of Calcium Silicate Hydrate (C–S–H) Near, Mid, and Far-Infrared spectroscopy." *J. Am. Ceram Soc,* 82 (3), 742–748.

CHAPTER 7

Freeze-Thaw Damage Reversal

DEGRADATION due to freezing and thawing is a problem for all po-rous roadbed materials, structural ceramics, and building façade materials. When frozen, fluid inside the pores of ceramic materials expands and develops tensile stresses which quickly lead to cracking. The first line of defense has, traditionally, been organic coatings that seal surfaces. Unfortunately, such coatings tend to be high in volatile organic content (VOC). In addition, surface treatments must, unfortunately, be perfect, or they admit water and allow it to settle more deeply into the wall. Here it still freezes, producing a more serious level of damage than it would have done closer to the surface. The alternative remedy, which we will examine here, is the use of reactive nanoparticles with other agents to fill pores more deeply. This method offers reduced porosity, healed microcracks, and a large increase in strength. The treatment used in this case relied upon electrophoresis and ionic conduction to transport pore-blocking particles and ions throughout the pore structure of a structural facade limestone (Cardenas *et al.* 2007). The following section describes the process and provides lessons that transfer easily for the benefit of concrete and other construction material applications.

7.1. FREEZE-THAW AND THE DISAPPOINTMENT OF SEALANTS

Limestone is a porous material, composed almost entirely of calcium carbonate. As with other construction ceramics, a major problem with limestone is degradation due to expansive freezing of absorbed moisture. The resultant cracking manifests as spalls and pop-outs, even craters. Damage proceeds deeper into interior sections, as they, in their turn, freeze and crack. Such damage, effectively increasing porosity, progressively degrades strength in stone work (Viklander and Eigenbrod, 2002). Furthermore, when in a state of thaw, the melted water moves inward to deeper and broader sections of the pore system, where it later refreezes. The result is the propagation of widespread damage.

There is no lack of coatings to seal the outer surfaces of masonry walls. They include polymer-based as well as ceramic species—some of them designed to stay at the surface, others to penetrate further. They may even induce a degree of strengthening within the outer millimeter or so of the stone (Brus and Kotlik, 1996; Kumar and Ginell, 1997). More significant, however, are the defects that almost always accompany mechanical application of surface barriers—and through which water can still migrate. Even if a perfect coating is laid down, the pores are not free of water that entered prior to the coating. In this case the water, residing on the interior side of the coating itself, is still close enough to the surface to permit freezing. More issues enter when dealing with organic sealants. High VOC contents create inconvenient, costly restrictions for disposal; and such products contribute to the production of ground-level ozone, which is associated with respiratory health risks. These issues encourage the use of a different option, one which provides protection at a greater depth within the wall while avoiding environmental drawbacks. The following sections describe how a nanoparticle treatment applied to structural limestone can achieve such outcomes. We will also explore some interesting material developments in the process that are applicable to concrete and other ceramics.

7.2. GETTING NANOPARTICLES INTO VERTICAL FAÇADE STRUCTURES

The application of treatments to vertical surfaces requires the use of a sponge electrode assembly, such as shown in Figure 7.1. Treatment fluids are fed onto the top side of each sponge, and gravity causes them to flow along the sponge. Rigid support holds the sponge against the surface, so that the fluid within it remains in contact with the wall surface. As each fluid travels down the wall, an electric field, applied using metal mesh electrodes, draws the nanoparticles and ions into the pores of the limestone. The carrier fluids are recycled to permit an adequate dosage of treatment to be delivered into the pores.

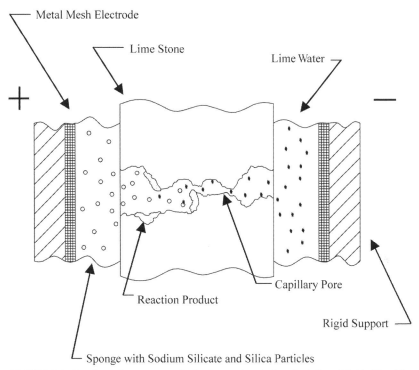

FIGURE 7.1. *Re-circulating sponge electrodes are located on the left and right side of the limestone block. The electric field pulls nanoparticles out of the fluid and into the pores of the wall. Adapted from Cardenas et al. © 2007 Taylor & Francis Group, London, UK. Used with permission.*

Two sizes of limestone block were used, 12.7-cm cubes and 5-cm cubes, obtained from Texas Quarries Inc., Cedar Park, TX. Cycles of freezing and thawing were carried out as per ASTM D 5312 (ASTM, 1992). During the freezing cycle, each specimen was kept in an individual container, each container partially filled with water. The temperature of the freezer was maintained in the range of –18 to –20°C during each 12-hour freeze cycle. Thawing cycles lasted 8 hours, the specimens maintained at 32°C. Each specimen experienced 10 full cycles of freeze and thaw.

Electrokinetic treatment was conducted for 12 days, using sodium silicate and calcium hydroxide solution. The sodium silicate was a 44 wt% formulation, obtained from Oxychem Inc., Dallas, TX. Reagent grade calcium hydroxide was combined with de-ionized water to produce a solution slightly exceeding the 1.7 g/l solubility limit. The limestone cubes were placed between two cellulose sponge electrodes, as illustrated in Figure 7.1. The electrodes were mixed-metal coated titanium wire mesh. A constant current density of 1 A per square meter was applied to each stone. In general, this required a voltage that ranged from a starting value of approximately 15 V to an ending value of approximately 30V. Four liters of each treatment fluid was circulated through the sponge electrodes during a 12-day treatment period.

Several treatment cases were conducted and replicated in 3 trials for each of the two sizes, the 5-cm cube, and the 12.7-cm cube. The cases are summarized as follows.

Case F-E: Freeze and thaw cycling followed by 12-day electrokinetic treatment.

Case E-F: Electrokinetic treatment for 12 days, followed by freeze and thaw cycling.

Case F: Freeze and thaw cycling only.

Case E: Electrokinetic treatment only.

Case C: Control specimens with no treatment or freezing.

7.3. A NEW FREEZE-THAW RESISTANT NANOCOMPOSITE

An SEM image of treated and untreated limestone is shown in Figure 7.2. The untreated sample, taken from the center of the block after undergoing 10 freeze and thaw cycles, shows an apparent microcrack. A broad range of grain sizes are evident, and the cracks observed appeared to pass between the larger and smaller grains. The image of the treated limestone shows a sample that was also taken from the center of the specimen. In this case, similar sizes of large particles are evi-

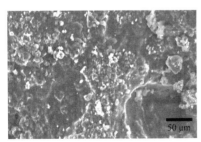

FIGURE 7.2. SEM image of specimens exposed to freeze and thaw cycling. Specimen on the left was protected by nanoparticle treatment. The image on the right, untreated, exhibits cracking due to freeze and thaw cycles. Adapted from Cardenas et al. © 2007 Taylor & Francis Group, London, UK. Used with permission.

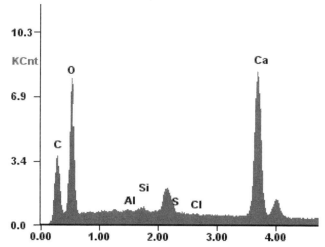

FIGURE 7.3. EDS spectrum of freeze and thaw cycled limestone specimen that had not been subjected to electrokinetic treatment. Reproduced from Cardenas et al. © 2007 Taylor & Francis Group, London, UK. Used with permission.

dent, but far fewer smaller particles can be seen. The visibility of the finer particles appears to be obscured by another, smoother phase, that appears to saturate much of the space between the larger- and medium-sized particles. There is thus little opportunity to distinguish the smaller particles prevalent in the untreated sample image.

Figure 7.3 shows an energy-dispersive spectroscopy (EDS) spectrum run on the untreated sample image of Figure 7.2. Carbon, oxygen and calcium were detected, along with small amounts of sulfur and chlorine.

Figure 7.4 plots the EDS spectra of the treated sample shown in Figure 7.2. Here again the spectrum shows the presence of carbon, oxygen, and calcium. Small amounts of sulfur and chlorine were observed again as well, and a strong peak for silicon was also prominent. The presence of silicon is due to the use of sodium silicate. The electric field inserted negatively charged silicates as well as colloidal silica particles of various sizes, and prevented sodium ions from entering the pores. This explains why no trace of sodium was detected in the EDS spectra of the treated sample.

Limestone specimens were subjected to chemical analysis to assess the amount of calcium carbonate present. The calcium carbonate contents are listed in Table 7.1.

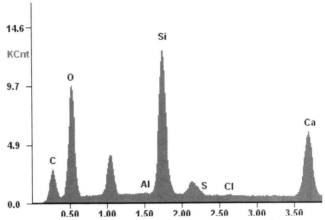

FIGURE 7.4. EDS spectrum of electrokinetically treated specimen from Figure 7.2. Reproduced from Cardenas et al. © 2007 Taylor & Francis Group, London, UK. Used with permission.

These values were provided by the Bowser-Morner Laboratories, Dayton, OH. There were two (un-treated) cases with calcium carbonate contents over 99%. The others that had received treatment exhibited calcium carbonate contents of 95–96%. This means that a composite material was formed that was 95% calcium carbonate with a 4% calcium silicate filler content. The remarkable strength contribution of this new phase will be examined in the next section.

7.4. GREAT COMPOSITE STRENGTH ENHANCEMENT

The bar chart in Figure 7.5 provides a comparison of strengths and volume porosities of the limestone-treatment cases. It was observed that undamaged and untreated cases exhibited a strength of 1,300 psi and a

TABLE 5.1. Calcium Carbonate Content.

Treatment Case	Calcium Carbonate (Weight %)
F-E	96.58
E-F	95.99
F	99.20
E	95.90
C	99.11

Reproduced from Cardenas *et al.* © 2007 Taylor & Francis Group, London, UK. Used with permission.

porosity of 10%. Electrokinetic treatment also provided a substantial strength increase for the undamaged specimens. This new strength was found to be 2800 psi. It was accompanied by a porosity that fell to 6%. When subjected to 10 freeze and thaw cycles, these specimens exhibited a significant drop in strength down to 880 psi and the porosity increased to 15%. When electrokinetic treatment was applied prior to freeze and thaw cycling, the strength climbed to 2,600 psi with a porosity value of 7%. When the treatment was used to recover properties after freeze and thaw exposure, the resultant strength was lower, but still impressive, at a value of 1,600 psi, with porosity at 9%—both still higher than the original, undamaged material.

In each case, the increase in strength was impressive in light of the small amount of new material injected. As noted earlier, the combination of calcium hydroxide and sodium silicate are known to form a variant of C–S–H. The new composite material may have as much as 4 volume percent of C–S–H. When formed in an electro-deposition mode, it is possible to produce a consolidated form of C–S–H that exhibits significant strength.

A crude estimate of the strength of this consolidated C–S–H phase is obtained using the composite rule of mixtures. This involves the assumption that we are dealing with reasonably linear-elastic materi-

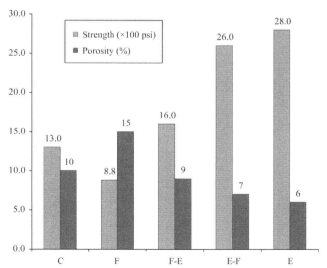

FIGURE 7.5. Comparison of strength and porosity for various treatment cases. Reproduced from Cardenas et al. © 2007 Taylor & Francis Group, London, UK. Used with permission.

als, and that fracture occurs in each phase at about the same level of strain. In this case, applying the composite rule of mixtures to the 90% calcium carbonate and 10% water (or air) content of the original lime stone yields a calculated strength in the solid calcium carbonate phase of 1300/0.9 or 1,444 psi. Now in examining the undamaged limestone that was treated to achieve a strength of 2800 psi, the application of the composite rule of mixtures can help elucidate the strength of the new C–S–H phase. The difference between the untreated and treated porosity was $10 - 6 = 4\%$. Applying the composite rule of mixtures gives a new phase strength of $[2800 - 1,444(0.9)] \div 0.04 = 37,500$ psi. This is a crude estimate; but from a historical perspective this value correlates to one of the highest strength (36,000 psi) concretes ever obtained with normal temperature and pressure curing (Hjorth, 1983). That study achieved this result using a highly consolidated combination of densified small particle (DSP) cement combined with microsilica. More recently, others have achieved a strength of 30,000 psi with a normal cure by using a combination of finely ground Portland cement powder, silica fume, glass powder, high-range water reducer, steel fibers, and a w/c ratio of 0.22 (Wille *et al.* 2011). This appears to indicate that the electrokinetic treatment produced a highly consolidated phase as the calcium ions and silicate species approached each other and piled up within the interior of the limestone. The properties of this consolidated calcium silicate appear to be at the same level strength as some ultra-high concrete compositions that do not use high temperature or pressure curing.

7.5. CRACKS vs. PORES

In general, strength is expected to decline as the porosity of a material increases (Cardenas, H., 2002 and Salem, N., 2003). This was so here. On a practical level, the damage caused by freezing and thawing resembles a simple increase in porosity. At the same time, however, it is clear that porosity and microcracking are very different things, because we are comparing an infinitely sharp crack with a stress concentration factor as high as 14 (or more), to a circular defect for which the stress concentration factor is only 3. It is likely that the growth of a microcrack would follow a relatively porous pathway, because this is the path of least resistance. A crack will arrest when the stress field at its tip falls below the stress intensity required for propagation—a type of event that will most likely occur at a pore. The end result of microcrack propagation is, then, a better connected pore network. Any damage obtained from freeze and thaw cycling may be manifest as a stepwise linking process between existing areas of porosity that already exhibit tensile cracking. Figure 7.6 appears to show a fairly linear relationship between strength and porosity, yet we know that in this case the porosity increases consist of the creation and propagation of cracks. For this reason, the accumulation of freeze and thaw damage may simply be considered a damage-related rise in porosity. With regard to damage healing, this distinction will be important, because filling a circular pore with a strengthening deposit is easier than inducing materials to travel down an infinitely sharp crack. It does not seem likely that the impressive strength increases observed among these treatment cases would be possible if large numbers of individual deposits had to be placed at the tips of infinitely sharp cracks.

As discussed in the previous section, it is conceivable that damage repair from microcracking can be observed in terms of a reduction in porosity. The porous paths that allow water to enter and then become increasingly interconnected as a result of freeze and thaw damage, also provide a pathway for restoration.

Similar results were observed in both the damaged and undamaged enhancement scenarios. Undamaged porous pathways decreased in volume from 10% to 6% due to treatment. In the same way, treatment of

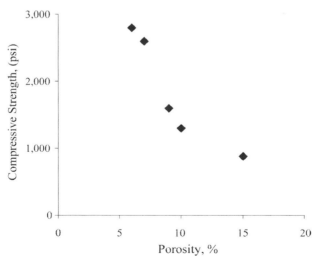

FIGURE 7.6. Relationship between altered porosity and strength. Reproduced from Cardenas et al. © 2007 Taylor & Francis Group, London, UK. Used with permission.

the damaged specimens reduced their apparent porosity values from 15% to 9%. The resultant strength increases for both undamaged and damaged cases were approximately 100%. It is interesting to note that the samples were harvested from the central interiors of each specimen. This means that the original undamaged pore system, as well as the microcrack-modified pore system, were effectively penetrated by restorative treatments with high impact and similar results.

7.6. THE DAMAGE RECOVERY TREATMENT

As shown in Figure 7.5, damaged specimens (F) showed much lower strength compared with undamaged controls (880 psi versus 1,300 psi), and significantly higher porosity (15% versus 10%). If treatment was applied following the application of freeze and thaw cycling (Case F-E), the damage-related porosity was reduced significantly from 15% back to 9%. This porosity value was comparable to the original porosity (10%). It was also observed that strength recovered and then exceeded the original undamaged material by a value up to 1,600 psi, or 23%. It appears from these observations that the electrokinetic treatment both restored and enhanced the damaged material, providing great strength and less porosity than the pre-damaged limestone.

7.7. THE PREVENTATIVE TREATMENT

Preventive treatments were applied prior to the freeze and thaw exposure. These treatments were designated as Case E-F. The intent was to prevent damage by blocking pores, thus inhibiting transport of moisture. These E-F cases exhibited strength values very much above the damaged material by a factor of ~3. They also exceeded the strength of the undamaged control specimens by a factor of ~2. The porosity value produced was 7% as compared to 10% for the control specimens, and 15% for the freeze-and-thaw-damaged but untreated stone. These observations show that the treatments greatly lowered the intensity of freeze and thaw damage by changing the pore structure.

7.8. SUMMARY OF TREATMENT FOR FREEZING AND THAWING RESISTANCE

It has been demonstrated that both the damaged and undamaged pathways (due to microcracks) were readily penetrated by treatments that allowed calcium hydroxide and sodium silicate to meet and react well inside the center of the limestone. This was a sustainable approach, more effective than surface treatment. Electrokinetic treatments restored and enhanced material properties of the damaged limestone, leaving it radically stronger and less porous than even the pre-damaged state, and reducing the susceptibility to freeze and thaw damage by blocking pores that enabled such degradation to occur. It is clear that freeze and thaw damage can be viewed simply as a damage-related rise in volume porosity. When it comes to healing damage, this distinction is significant, because depositing material into a circular pore is a lot easier than filling the tip of an infinitely sharp crack.

Compared to restorative treatments, preventive treatments were much more effective at reducing damage. The deep penetration achieved by electrokinetics provided protective inorganic sealants that minimized the possibility of sub-surface freezing.

7.9. REFERENCES

ASTM (1992) D 5312-92. *Standard Test Method for Evaluation of Durability of Rock for Erosion Control Under Freezing and Thawing Conditions.* American Society For Testing Materials: Philadelphia.

Brus, J. and Kotlik, P. (1996) "Consolidation of Stone by Mixtures of Alkoxysilane and Acrylic Polymer," *Studies in Conservation,* Vol. 41, No. 2, pp. 109–119.

Cardenas H. (2002) *Investigation of Reactive Electrokinetic Processes for Permeability Reduction in Hardened Cement Paste,* PhD Thesis, University of Illinois at Urbana-Champaign, pp.1.

Cardenas, H., Paturi, P., Dubasi, P., "Electrokinetic Treatment for Freezing and Thawing Damage Mitigation within Limestone", *Proceedings of Sustainable Construction Materials and Technologies,* Coventry England, Taylor & Francis Group, UK 12 June 2007.

Hjorth, L., "Development and Application of High-Density Cement- Based Materials," *Philosophical Transactions of the Royal Society of London,* A 310, pp. 167–173, 1983.

Kumar, R. and Ginell, W. S. (1997) "A New Technique for Depth of Penetration of Consolidants into Limestone Using Iodine Vapor," *Journal of the American Institute of Conservation,* Vol. 36, No. 2, pp. 143–50.

Salem, N. "Effect of Porosity on Properties of Concrete," *Journal of Hungarian Group of Concrete Structures.* http://fib.bme.hu/fib/cikk/v03_eng_full/cikk03-8.php3, Budapest Hungary, 2003.

Viklander, P. and Eigenbrod, D. (2000) "Stone Movements and Permeability Changes in Till Caused by Freezing and Thawing," *Cold Regions Science and Technology,* Vol. 31, No. 2, pp. 151–162.

Wille, K., Naaman, A., Parra-Montesinos, G., "Ultra-High Performance Concrete with Compressive Strength Exceeding 150 MPa (22 ksi): A Simpler Way", *ACI Materials Journal of the American Concrete Institute,* Jan/Feb 2011.

CHAPTER 8

Electrokinetic Nanomaterial Process Control and Design

A number of phenomena can influence the location and transport of nanoparticles. Some of them produce independent motions in more than one direction, each with the same exact driving force. The following sections examine basic principles of transport control and the development of a model that predicts the rate of nanoparticle transport within concrete. The final section takes the reader through a design example.

8.1. TRANSPORT CONTROL

Concrete, as a porous material, exhibits durability largely governed by the transport of chemical species such as sulfate, chloride, and carbon dioxide. In general, transport in such a porous material involves a potential difference that causes several related transport processes or fluxes.

8.1.1. General Flow Laws

Potential differences include pressure, temperature, electrical, and chemical potential gradients. The fundamental flows influenced by these gradients are those of bulk fluid, heat, electric current, and chemical species. A given gradient can drive more than one of these flows at the same time. In a given case, a particular flow may be favored and easily observed, while others may not be detectable. The favored flows and their respective gradients are linearly related via conductivity coefficients of the general form (Mitchell, 1976):

$$J_i = L_{ii}X_i \qquad (8.1)$$

where J_i is a measured flow, L_{ii} is a conductivity constant pertaining to the specific flow, and X_i is the primary potential gradient influencing the flow.

In the case of a constant electrical gradient, one example expression is known as Ohms Law:

$$I = \sigma\Delta E \qquad (8.2)$$

where I is the electric current, σ is electrical conductivity, and ΔE is electrical potential difference.

A given potential gradient may cause the development of other gradients leading to coupled flows. The influence of various gradients on a particular flow is described by the following general expression:

$$J_i = L_{ij}X_i \qquad (8.3)$$

where L_{ij} is the matrix of transport coupling coefficients, and X_i is the array of potential gradients.

In a general case, the expression of water flow J_H under the influence of the major potential gradients is given by:

$$J_H = L_{HP}X_P + L_{HT}X_T + L_{HE}X_E + L_{HC}X_C \qquad (8.4)$$

where HP is a hydraulic/pressure component, P is a pressure component, HT is a thermal/hydraulic coupled component, T is a thermal component, HE is an electrical/hydraulic or electroosmotic component, E is an electrical component, HC is a chemical/hydraulic or normal osmotic component, and C is a chemical component.

The relative contribution of a given gradient is governed by the physical and chemical characteristics of the environment under consideration. Each of the singular and coupled coefficients is typically generated empirically as needed for specific cases. Many versions of these coefficients have also been analytically developed for specific cases. In some cases, such as hydraulic flows in cementitious materials, singular permeability coefficients have been developed that are partially analytical, enabling some of the particular environmental influences to be physically characterized.

8.2. TRANSPORT PHENOMENA IN CONCRETE

The preceding sections relate primarily to saturated porous materials. A number of distinct transport processes involving both saturated and unsaturated concrete have been described in the literature. These processes include adsorption, liquid diffusion, capillary absorption, bulk laminar flow, ionic conduction, electrophoresis, and electroosmosis (Hearn, Hooton and Mills, 1994). In general, each of these processes is governed by the geometric and chemical properties of the concrete pore system as well as the water content of the pores. The following sections contain descriptions of the characteristics of each process.

8.2.1. Adsorption

At relative humidities below 45%, the hydrophilic nature of cement leads to significant adsorption of water on the pore walls (Powers, 1960). This process depends on the local surface energy at any point along a given pore wall. At relative humidities above 45%, the rate of adsorption is reduced as the number of available sites for water vapor becomes limited. The surface energy is such that adsorption continues to take place after the initial monolayer is in place, leading to multi-layer adsorption.

8.2.2. Diffusion

The basic driving force in diffusion is the reduction of energy achieved when a chemical constituent diffuses to a more uniform concentration. This movement is in the direction of lower concentration. Fick's first law relates the mass transfer rate to the concentration gradient (assuming steady state conditions) as follows:

$$Q_m = D_c \frac{dc}{dx} \tag{8.5}$$

where Q_m is the mass transport rate, (mol/m^{2}s), D_c is the diffusion coefficient (m^2/s), and dc/dx is the concentration gradient (mol/m^3/m) (Ashby and Jones, 1980). This expression has the form of a general flow expression, as examined in earlier sections. In this case the primary coefficient is a chemical component.

Fick's second law provides a solution that governs the non-steady state cases, which has the form (Hearn, Hooton and Mills 1994):

$$C(x,t) = C_c \left[1 - erf \left\{ \frac{x}{2} \sqrt{D_c} \right\} \cdot 1 \right] \tag{8.6}$$

where $C(x,t)$ is the ionic species concentration at location x and time t, C_c is the surface concentration of ionic species, and erf is the Gaussian error function (values are given in standard tables). This expression is valid for a one-dimensional semi-infinite case.

Recent work dealing with chlorides in cement has re-examined the assumption that the diffusion coefficient (D_c) is dependent upon the concentration gradient alone. Empirical evidence has shown that the diffusion coefficient is a variable that is sensitive to the absolute species concentration at a given location in the pore (Chatterji, 1999). An expression for this variable diffusion coefficient in porous media is given as:

$$D_c = D_o - k_c C^N \tag{8.7}$$

where D_o, D_c are the diffusivities representative of zero ion concentration and concentration C, k_c is a pore characteristics parameter, and N is an empirical constant.

Additional refinement is obtained by considering the influence of other species and their respective directions of travel. These interactions have been formulated in terms of a diffusion potential averaged over a representative control volume, as well as through application of an expanded version of the Nernst equation (Barbarulo *et al.*, 2000 and Samson *et al.*, 2000).

Diffusion in concrete depends upon the amount and composition of the fluid in the pores. The following sections describe the contributing roles of liquid and vapor diffusion in concrete.

8.2.3. Surface Diffusion

In the case of water vapor, adsorption onto pore surfaces leads to the development of a multi-layered film that exhibits a specific gravity of approximately 0.9 and a very high viscosity. The properties of this film have been compared to a Type IV ice (Hearn, Hooton and Mills,

1994). The observation of surface diffusion along this film is associated with elevated vapor diffusion rates of adsorbing gases (such as water) as compared to the diffusion of non-adsorbing gases such as nitrogen.

8.2.4. Vapor and Liquid Diffusion

The processes of water vapor diffusion, surface adsorption, and surface liquid diffusion lead to the development of menisci. For a given molecule, the transport process may involve a complicated series of phase transitions governed by local concentration gradients in gas and liquid phases, and by local surface energies that influence film properties along pore surfaces. Under the influence of elevated vapor pressure, a molecule in the vapor state may enter the liquid state at an upstream meniscus. The change in energy within the bulk fluid leads to the evaporation of a water particle (for example) from a downstream meniscus.

8.2.5. Capillary Absorption

This powerful transport process is driven by the reduction in surface energy of the pore as the bulk liquid wets and spreads over the pore wall. Key ingredients for capillary absorption flow are the presence of unsaturated surfaces and sufficient water to form menisci. This is a dominant transport process even when applied pressure is influencing the flow (Reinhardt, 1992).

8.2.6. Hydraulic Flow Permeability

As noted earlier in the permeability section, the application of applied pressure causes the development of laminar flow in the concrete pores as described by Darcy's law [Equation (2.1)]. An alternative formulation of Darcy's law in terms of volume flow rate may be written:

$$Q_p = KiA \tag{8.8}$$

where A is the cross-sectional area of the flow path and i is the ratio of water head to the length of the flow path. For incompressible flows, the law of conservation of mass reduces Equation (8.8) to the continuity equation, which has the form:

$$Q_p = V_{p1}A_{p1} = V_{p2}A_{p2} = \text{constant} \tag{8.9}$$

where Q_p is the hydraulic volume flow rate, V_{p1}, V_{p2} are the flow velocities, and A_{p1}, A_{p2} are the cross-sectional areas, both at arbitrary locations 1 and 2. Inserting the expressions for hydraulic gradient and continuity into Darcy's law yields the useful form:

$$Q_p = K\frac{h}{L}A \qquad (8.10)$$

As noted in the section on permeability reduction, the coefficient K has been modeled in other empirical ways (Garboczi, 1990). A model for K in terms of pore structure variables is given by:

$$K = d_c^2\phi\beta \qquad (8.11)$$

where d_c is the threshold pore diameter based on capillary pore size distribution obtained via mercury intrusion porosimetry (MIP), ø is the volume fraction of pore space, and β is the parameter for tortuosity and connectivity.

Kozeny and Carman theory for transport was to treat porous structures as a collection of parallel tubes of circular cross-section. For permeability applications it was adapted to provide the following model (Garboczi, 1990; Carman, 1956):

$$K = \frac{\phi^3}{2S_s^2} \qquad (8.12)$$

where S_s is the specific surface area. In principle, this model treats the pore structure as a series of tubes. In the application of it, uncertainty emerges from the difficulty of obtaining an accurate surface area measurement.

The Katz-Thompson model, formulated in terms of a relative conductivity parameter, is given by:

$$K = \frac{C_k d_c^2}{F} \qquad (8.13)$$

where C_k is a calculated constant (1/226) and F is the formation factor, defined as σ^o/σ where σ^o is the conductivity of pore water and σ is the conductivity of the bulk material. The preceding flow models yield good results if successive values are within an order of magnitude in agreement.

One method of characterizing pore sizes and determining d_c is through mercury intrusion porosimetry (MIP). In this method, applied pressure is used to intrude mercury into the pores of the sample. The volume of mercury intruded into the paste is related to the pore size being entered, the applied pressure, and the surface tension of mercury. This relationship is governed by the Washburn equation. Some limitations regarding both true and measurable pore sizes, as well as microstructural damage, are significant issues (Cook and Hover, 1999; Taylor, 1997, p. 249; Garboczi and Bentz, 1991; and Diamond, 2000).

The permeability models mentioned earlier in this section work best when capillary forces are not significant compared to the applied pressure. For the case where capillary forces are significant with respect to hydraulic forces, a useful one-dimensional flow expression for a wetting front is given by (Reinhardt, 1992):

$$x(t) = C_p t^{1/2} \tag{8.14}$$

where the $x(t)$ is the time dependent location of the wetting front, C_P is the fluid penetration coefficient, and t is time. If capillary and hydraulic forces are significant, then (Reinhardt, 1992):

$$C_P = \frac{r}{2} \left[\frac{P_e + P_c}{\eta} \right]^{1/2} \tag{8.15}$$

where P_e is the hydraulic pressure, P_c is the capillary tension, η is the dynamic viscosity of the liquid, and r is the radius of the cylindrical capillary. The impact of P_e may be neglected if it is less than 1 meter of head pressure.

8.3. ELECTROKINETIC TRANSPORT

Porous systems such as cement paste and concrete may play host to various electrokinetic phenomena, including ionic conduction, electrophoresis, and electroosmosis. The following sections discuss the observations and modeling associated with these phenomena.

8.3.1. Ionic Conduction

Ionic conduction accounts for the overwhelming majority of conductivity measured in cement-based materials. Monitoring the conductivity of cement pastes has gained acceptance in evaluating various hydration phenomena, such as changes in pore structure and transformation of phases (Morsy, 1999). Examination of aqueous ionic conduction establishes a fundamental building block for analysis of various electrokinetic behaviors.

In an aqueous system, ions can be induced to drift in response to an applied electric field. During this process, a number of forces influence motion. When no net drift is occurring (prior to the application of an electric field, or some other perturbation), a spherically symmetrical ion cloud exists around the central ion. If the central ion exhibits a positive charge, the cloud consists of negatively charged species, randomly oriented water molecules, and water molecules oriented by the electric field of the central ion. When an electric field is applied, the ion cloud takes on an egg shape as species are added to the cloud on the leading side in the direction of motion, and others are left behind on the trailing side (Bockris and Reddy, 1976, p. 424). The term relaxation field has been applied to this egg-shaped cloud, because this shape is associated with the continual decay or relaxation of charge on the trailing side. During this process, the center of charge in the ion cloud no longer coincides with the center of charge of the central ion. This condition is referred to as the relaxation field force (F_r), which acts in opposition to the drift velocity of the central ion.

An additional force working against the transport of the central ion is associated with the electric field acting on the ion cloud. Since a given ion cloud can be on the order of 10^{-8} meters in diameter, its influence has been referred to as the electrophoretic retardation force (F_e). As with all objects in motion in a fluid medium, the central ion is subject to a viscous drag force, which Stokes and Robinson (1955) derived in the form:

$$F_d = 6\eta \upsilon \pi r \tag{8.16}$$

where r is the radius of the ion cloud, η is the viscosity of the bulk fluid, and is the drift velocity of the central ion.

The relaxation force mentioned earlier is given by:

$$F_r = \frac{z^3 e_o^3 r X_E}{D K_b T} \tag{8.17}$$

where r is the radius of the ion cloud, X_E is the applied electric field, z is the valence of the central ion, e_o is the unit charge in coulombs, D is the dielectric constant of the medium, K_b is Boltzmann's constant, and T is the absolute temperature.

The electrophoretic retardation force is given by:

$$F_r = \Sigma z_i e_o X_e \tag{8.18}$$

where a summation on i is assumed for the collection of charges in the ion cloud. Meanwhile, the driving force for the conduction of the central ion is given by:

$$F_c = z e_o X_e \tag{8.19}$$

Now that each of the major force components have been identified, the net quasi-steady state drift velocity is given by:

$$F_c = F_e + F_d + F_r \tag{8.20}$$

Inserting Equation (8.16) into this expression provides a solution for the drift velocity of an ion. A conductivity measurement value is proportional to the average drift velocity of the collection of ions in a test specimen.

8.3.2. Electrophoresis

This phenomenon is characterized by the movement of a solid particle dispersed in an electrolyte under the influence of an electric field. A key aspect of this behavior is the presence of a surface double layer of charge on the solid particle comprised of ions adsorbed and attracted to the surface (Bockris and Reddy, 1976, p. 742). Clay, cement, and silica particles tend to carry a negative charge. The fixed portion of the double layer on these particles consists largely of positive charges adsorbed to the surface. Beyond the adsorbed layer is a layer that consists of a diffuse concentration of positive charges that are also attracted to the surface of the negatively charged particle. Negative ions are also present in this layer, because they are attracted to the concentration of positive charge. The diffuse and mobile nature of this outer layer significantly influences the electrokinetic mobility of the particle. A potential difference exists between the inner boundary of the diffuse layer and the bulk fluid. This potential is referred to as the electrokinetic or zeta potential (Z). The diffuse collection of charges in the double layer has a similar role in electrophoresis, as discussed in the previous section on ionic conduction, with similar influences stemming from viscous, relaxation, and electrophoretic forces.

An estimation of the electrophoretic mobility is obtained by assuming a particle of spherical shape. The potential at the surface of the particle is related to the outer layer charge q_d via:

$$q_d = ZrD \qquad (8.21)$$

Meanwhile, the electric driving force operating on the particle plus cloud is given by:

$$F = X_E q_d \qquad (8.22)$$

Eliminating q_d from the preceding two expressions, and setting the driving force equal to the Stoke's viscous drag force [Equation (8.16)] yields:

$$\upsilon = \frac{X_E ZD}{6\eta\pi} \qquad (8.23)$$

where, Z in this expression is the zeta potential of the double layer. This expression yields an estimate of the electrophoretic velocity, neglecting the relaxation and electrophoretic retardation forces associated with the double layer.

8.3.3. Electroosmosis

In 1809, Ruess observed that application of a potential difference across a clay or sand system caused water to flow from one electrode to the other (Glasstone, 1946, p. 1220). The mode of transport was found to resemble plug flow, which refers to flow with a relatively uniform cross-sectional velocity profile (Bockris and Reddy, 1976, p.826). The application of an electrical potential and pressure each contribute to the observed flow in accordance with the following relation:

$$V = a_1 \Delta P + a_2 \Delta E \qquad (8.24)$$

where V is the fluid flow velocity, a_1 and a_2 are empirical coefficients, ΔP is the applied pressure drop, and ΔE is the applied potential difference. Note that this expression has the form of the generalized coupled flow law discussed in Section 8.1.1. A key point here that is expressed in Equation (8.24)—that a given pressure flow can be slowed, stopped, or even reversed by the application of an electrical potential. Reversing the polarity of this potential would cause the flow to increase in magnitude.

Either an applied pressure and/or an applied potential causes an electrical current to flow in accordance with the following relation:

$$I = a_3 \Delta P + a_4 \Delta E \qquad (8.25)$$

where a_3 and a_4 are empirical coefficients. The Onsager reciprocity relation applied to Equations (8.24) and (8.25) yields the following cross-coefficient correlation:

$$(I/\Delta P)_{\Delta E} = 0 = a_3 = a_2 = (V_{eo}/\Delta E)_{\Delta P} = 0 \qquad (8.26)$$

where V_{eo} is the electroosmotic velocity. This expression essentially states that the electric current per unit of applied pressure is equivalent to the fluid velocity per unit of applied potential. The expression has been confirmed experimentally.

The application of a hydraulic pressure difference causes the development of a potential difference and an electrical current. These responses are referred to as the streaming potential and the streaming current, and are governed by the preceding reciprocity relation.

The source of all these phenomena is related to the electrical double layer that is associated with electrophoresis. As shown in Figure 8.1, the double layer consists of a fixed layer of charges and a shear layer of charges. Outside the shear layer is a diffuse region of charges of relatively lower concentration. An applied electric field causes the ions in the shear layer to flow toward one of the electrodes. This motion

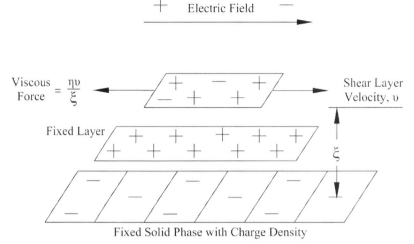

FIGURE 8.1. *Origin of viscous force responsible for plug flow in electroosmosis. (Adapted from Bockris and Reddy, 1976, p. 830).*

imparts a viscous force onto the neighboring bulk liquid, generating the plug flow that carries with it the entire shear layer and the diffuse region of the pore fluid (Hunter, 1992; Bockris and Reddy, 1976; and Glasstone, 1946).

In the case of cement paste, the pore wall carries a negative charge. The origin of this negative surface charge stems from the nanostructure of the hydrated paste surface. Figure 8.2 contains a schematic representation of the silicate chain associated with C–S–H type materials such as 1.4 nm tobermorite and jennite (Taylor, 1997, p.135). The far left row of oxygen atoms in Figure 8.2 is part of the central CaO_2 layer. The atoms on the right-hand side of the structure are associated with the pore wall surface. Note that some of the oxygen atoms, lacking the hydrogen atom, carry a negative charge. The reaction by which these hydroxyl groups dissociate is given by:

$$SiOH \rightarrow SiO^- + H^+ \qquad (8.27)$$

This tendency to dissociation is sensitive to pH (Mitchell, 1976, p.126). Higher pH leads to more hydrogen ion dissolution and thus to a higher net negative charge carried by the material surface. In hardened cement paste, the pH values are in the range of 13–14, so it is likely that a strongly negative surface charge is present.

Pore fluid cations such as solvated calcium are attracted to these negative surface charges and form the Stern (fixed) layer (Shaw, p. 182, 1992, and Marchand *et al.*, 1998, p. 153). In electroosmosis, the double layer can be schematically rendered as shown in Figure 8.2. The immobile solid phase is the pore wall. A fixed layer of positive charges is located adjacent to the pore wall. Outside this fixed layer is a diffuse layer with a net-positive distribution of ions. When a potential field is applied along the pore, the positive ions in it will drift in the direction of the applied electrical current. The motion of this diffuse (shear) layer applies a drag force on the bulk fluid region, producing electroosmotic flow.

In typical cases where the pH of the hardened cement is elevated, or where there is a significant concentration of anions, the double layer has been described as a modified Stern Double Layer for hardened cement paste (Zhang *et al.*, 2001 and Marchand *et al.*, 1998). In this case, the Stern layer was conceived as two layers with positive ions (mostly potassium, sodium and calcium) and water dipoles on the inner layer (close to the pore wall). The outer Stern layer contains negatively charged ions and water dipoles. The net charge of these two layers is positive.

The potential difference (y) between the pore wall and the surface of shear is referred to as the zeta potential (Z). An expression for the electroosmotic velocity is given by:

$$\frac{V_{eo}}{X_E} = \frac{ZD}{4\pi\eta} \tag{8.28}$$

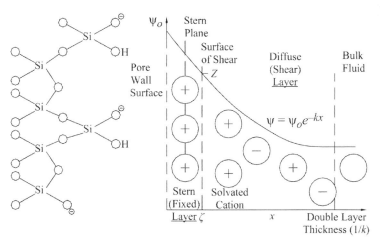

FIGURE 8.2. Nanostructure and charge distribution in C–S–H/pore fluid interface. (Adapted from Taylor, 1997, p. 135, and from Shaw, 1992, p. 183).

This model was presented by Helmholtz in 1879 and refined to the current form by Smoluchowski in 1914 (Mitchell, 1976). The degree of electroosmotic effect depends on the level of adhesion of the diffuse (shear) layer to the pore wall. The concentration of the solution is also a significant factor. In highly concentrated solutions, the electroosmotic flow phenomenon is not observed because the ions in the bulk fluid invade the double layer, causing the potential or the double layer thickness ($1/k$), or both, to be reduced. Both electroosmotic flow and zeta potential are sensitive to changes in electrolyte chemistry, which has been known to change the sign of the zeta potential. Figure 8.2 represents a double layer speciation that exhibits a negative zeta potential. Changes to the speciation of the pore fluid could raise or lower the zeta and possibly cause it to change sign. It is conceivable that the rate, and possibly the direction, of electroosmosis could vary with the age and alkali content of the cement paste or the inclusion of a surfactant.

The Helmholtz-Smoluchowski theory for electroosmosis is considered to be a large-pore theory. In the early 1950's, Schmidt put forth the following small-pore theory for electroosmosis (Mitchell, 1976, p. 357):

$$Q_A = \frac{A_o F_o r^2 N_P X_E A}{8\eta} \tag{8.29}$$

where A_o is the concentration of wall charges in ionic equivalents per unit volume of pore fluid, F_o is Faradays constant, r is the pore radius, N_P is the number of pores in the sample, X_E is the electric field, A is the cross-sectional area of the sample, and η is the viscosity. Note the second order dependence on pore radius. This expression is considered to be a small-pore theory and does not agree well with large-pore behavior.

A key aspect of observing electroosmosis in cement is the use of a high water/cement ratio. Workers have experienced frustration in attempting to demonstrate electroosmotic flows while using w/c ratios in the vicinity of 0.5 (Hayes, 1997). The reason for this difficulty is that, according to the Schmid theory for electroosmosis, the volume flow rate is a second order function of the pore radius (Mitchell, 1976, p. 271). The pore radii in cement paste may vary over 3 orders of magnitude. When a low-quality cement paste with a w/c ratio of 0.8 exhibits an electroosmotic coefficient of permeability (K_{eo}) at the low end of most observable cases it is not surprising that an order of magnitude drop in pore radii may yield a K_{eo} reduced by 2 orders of magnitude. Such a reduction would yield a K_{eo} of 0.003 ml/hour.

8.4. NANOPARTICLE TRANSPORT MODELING IN CONCRETE

In this section, a model is described that predicts the rate of nanoparticle penetration into concrete. In it, the principle of superposition is used to manage the impacts of electrophoresis, electroosmosis, and hydraulic flow. The model works by finding the penetration depth. This is done by dividing the net particle velocity by the number of degrees of freedom it has as it passes through the pore system. The result is then multiplied by the treatment time in order to obtain the penetration depth. The model compares well with experimental results, both for predicting penetration as well as the time it takes to achieve it.

8.4.1. Model Velocity Components

When Darcy's law is reduced to a velocity, as opposed to a volume flow rate, both sides of Equation (8.8) must be divided by the cross-sectional area of the flow. The law then has the form:

$$v = \frac{K \cdot h}{L} \tag{8.30}$$

where h is the pressure head, and L is the nominal length through the porous material (Darcy, 1856).

The flow travels through pores. This means that v may be divided by the porosity, in order to estimate the average velocity through the pores as follows:

$$V_P = \frac{K \cdot h}{L \cdot \phi} \tag{8.31}$$

where V_P is the average velocity in the pores. The porosity ø is usually less than 0.4. Based on conservation of momentum, it is assumed that the velocity of water in the pores is larger than at locations where it is either approaching the pore entrance or just leaving the pore system at the other end of the concrete specimen.

Electroosmosis contributes a bulk fluid velocity that is governed by:

$$V_{eo} = \frac{A_o F_o r^2 \Delta E}{8 \eta l} \qquad (8.32)$$

where A_o is the concentration of wall charges, F_o is Faraday's constant, r is the radius of the average pore, l is the length of the average pore, η is the absolute viscosity, and ΔE is the applied voltage (Mitchell, 1976, p. 357). In a complex pore system, the path-specific terms are consolidated together into a coefficient reducing the expression to:

$$V_{eo} = \frac{K_{eo} \Delta E}{L \cdot \phi} \qquad (8.33)$$

where K_{eo} is the electroosmotic coefficient of permeability. The porosity ø is used here again to obtain the average velocity within a pore as before.

Electrophoretic velocity is given by:

$$V_{ep} = \frac{Z \cdot D \cdot \Delta E}{6 \pi \cdot \eta \cdot L} \qquad (8.34)$$

where Z is the zeta potential and D is the dielectric constant (Bockris and Reddy, 1976, p. 834). The term V_{ep} is defined as the electrophoretic mobility, with units of m/s per unit of the applied electric field, $\Delta E/L$.

Now that the three main velocity terms have been introduced, we can turn to assembling them into a practical model. To do so, we will need to modify these terms into quantities that are easily measured.

8.4.2. Model Assembly

A transport model for practical use can be arranged to provide the distance to be traveled or the time it takes to make the transit. In this case, we will arrange it to give the depth of penetration, since this is a fundamental parameter that is easy to use for either predicting or interpreting the effectiveness of a given treatment. When a colloidal suspension is moved into the pores of concrete, the volume flow rate of the nanoparticles is expected to govern the depth of transport. The following sections describe the assembly of this model for predicting penetration depth.

8.4.2.1. *Theoretical Expression for Penetration*

In order to gain entry to a pore, the nanoparticle velocity must be greater than both the hydraulic pressure flow and the electroosmotic flow that oppose it. The following is an approximate expression for estimating a given nanoparticle treatment penetration:

$$P = \frac{V_{particle} \cdot t}{N_D} \tag{8.35}$$

where P is the penetration depth, V is the net particle velocity, t the total time that the treatment is being applied, and N_D the number of degrees of freedom of the porous pathway. The typical capillary pore is assumed to have a path that exhibits 5 degrees of freedom. These 5 degrees are the 5 directions of up, down, left, right, and forward. Backward direction of travel is excluded from the count only because it is not likely to play a significant role in establishing the average depth of penetration detected by examination of a fracture surface. The reason is that the longer paths of travel are associated with going backward. This will keep any particles that happen to travel backward from achieving the same depth as the rest of the particles during the same travel time. The greatest penetration will be obtained by nanoparticles whose paths do not involve backward sections, and it is these leading particles that will contribute to the observed penetration front seen on a fracture surface.

8.4.2.2. *Transport Control Volume*

A control volume is a theoretical construction that helps define the boundary conditions and terms needed to model a transport process. Figure 8.3 contains a control volume that represents a capillary pore undergoing multi-phase transport. All motion is being driven by the electric field, X_E, and the pressure gradient, X_P. These are force fields that induce electroosmotic (V_{eo}) and hydraulic (V_p) flows in the bulk

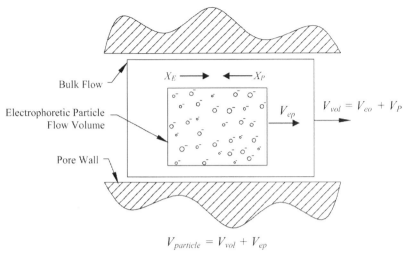

FIGURE 8.3. *Reactive electrophoretic transport control volume model.*

fluid contained within the control volume. Another control volume has been defined within the bulk fluid control volume. This smaller control volume is defined for just the charged nanoparticles being subjected to the electric field. The construction of control volumes helps to clarify the complex relationships between the driving forces and the motion of the different phases of matter. It also enables the visualization of a net particle velocity, $V_{particle}$, which is obtained by the superposition-based expression of Equation (8.36).

It is given by:

$$V_{particle} = V_{vol} + V_{ep} = V_{eo} + V_p + V_{ep} \qquad (8.36)$$

On a practical level, it is relatively simple to measure a volume flow rate as compared to the absolute velocity of a given particle or flow. The simple relationship between flow rates of velocity and volume is:

$$Q = V \cdot A \qquad (8.37)$$

where Q is the volume flow rate, V is the absolute velocity, and A is the treated surface area.

A key parameter in the electrophoretic component of the flow is the volume fraction of the suspension occupied by particles, f_v. This parameter is a calculated dimensionless term that will be defined in a later section. It is used here in the electrophoretic component of the volume flow rate as follows:

$$Q_{ep} = V_{ep} \cdot A \cdot f_v \tag{8.38}$$

It is interesting to note that this equation treats the volume flow rate of particles as if they all occupied a homogeneous solid volume. This odd picture actually correlates to the first stage of particle interaction, with the pore fluid in concrete, in which flocculation and precipitation occurs. As particles enter a pore, the high-ion content of the pore fluid forces them to start flocking and precipitating out of suspension. The precipitate takes up some of the ions in the vicinity, thus providing a localized depletion of both nanoparticles and ions native to the pore fluid. The next group of nanoparticles that drift into the pores find it depleted in ion content. This means that they are able to get further into the pore before they finally precipitate. Figure 8.4 illustrates this process, which is described as a traveling precipitation front. Using f_v makes it convenient to crudely approximate the aggregation of the nanoparticles once they enter the pores. It takes all the nanoparticles that are suspended and collects them into one solid mass that is assumed to completely fill initial sections of each pore prior to achieving further penetration. To be sure, this constraint is not physically correct, but its mathematical simplification describes an end result similar to the real outcome. This

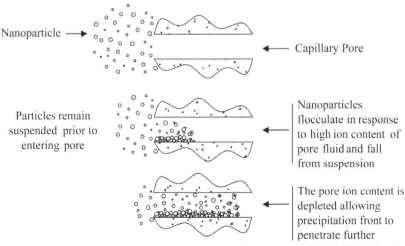

FIGURE 8.4. *Illustration of a traveling precipitation front theorized for the filling of capillary pores. Reproduced from Cardenas and Struble, 2008, with permission from ASCE.*

constraint is conservative. A real traveling precipitation front would probably allow somewhat more penetration, because the packing efficiency of nanoparticles permits a porosity reduction that is, typically, only around 30% in a bulk material treatment. This means that the particles are spreading out more that the model assumes.

Because they each represent bulk volume flows of fluid, modeling statements for the hydraulic and electroosmotic flows are convenient to handle together. As we take this step, it is helpful to relate the volume flow rates that exist within the concrete pores to the actual velocities within the pores. The porosity of the pore system permits us to relate these terms as follows:

$$Q_P + Q_{eo} = (V_P + V_{eo}) \cdot A \cdot \varnothing \tag{8.39}$$

where $\varnothing$ is the total capillary pore volume (porosity) of the concrete.

8.4.2.3. Nanoparticle Volume Concentration

The particle volume concentration, f_v, is generally not available from a nanoparticle manufacturer, but it can be determined from standard information that is provided. It is usually easy to obtain the density and the mass ratio of the nanoparticle content. These can be used, in the following expression, for the density of a *sol* that enables us to easily isolate and solve for the particle volume concentration. This expression is given by:

$$d_{sol} = \frac{d_w \cdot V_w + d_p \cdot V_S}{V_w \cdot V_S} \tag{8.40}$$

where d_{sol} is the density of the *sol*, d_w is the density of water, V_w is the volume of the sol occupied by water, d_p is the density of the sol particle, and V_S is the volume of the sol occupied by particles. Another vendor-supplied quantity is the mass ratio of particle content. Analytically, this is given by:

$$R_m = \frac{d_p \cdot V_S}{d_w \cdot V_w + d_p \cdot V_S} \tag{8.41}$$

The following expression is the volume of the sol occupied by particles. It is given by:

$$V_S = f_v \cdot U \tag{8.42}$$

where f_v is the volume fraction occupied by the sol particles and U is the total volume of the sol. A simple corollary for the water content in terms of V_W is given by:

$$V_W = (1 - f_v) \cdot U \tag{8.43}$$

At this point we have four equations containing five unknowns: f_v, U, V_S, V_w, and d_p. As an example, we can use Nalco 1050 colloidal silica, which has the values: $d_{sol} = 1.39$ g/ml and $R_m = 0.5$. Using these values in these equations, while eliminating V_S and V_W from Equations (8.40) and (8.43), permits U to cancel out. Now we have two equations with two unknowns that we can solve. This leaves us with the nanoparticle volume concentration for Nalco 1050 colloidal silica as $f_v = 0.31[(cm^3 - particle)/(ml - sol)]$.

8.4.2.4. Functional Expression for Transport

The final assembly of the penetration transport model starts by using the primary velocity terms identified in Figure 8.3. These velocities are now superimposed to provide a revised version of Equation (8.35):

$$P = \frac{(V_{eo} + V_P + V_{ep}) \cdot t}{N_D} \tag{8.44}$$

Here we have a first order expression for the rate of treatment penetration into concrete during a reactive electrokinetic nanoparticle transport process, in which both hydraulic and electroosmotic processes are playing a significant role. From a practical standpoint, it is highly convenient to redefine this penetration expression using volume flow rates instead of absolute velocities. We can do so by inserting Equations (8.38) and (8.39) into Equation (8.44). This eliminates the absolute velocity terms and provides us with:

$$P = \frac{\left(\dfrac{Q_{eo} + Q_p}{\phi} + \dfrac{Q_{ep}}{f_v} \right)}{A \cdot N_D} \cdot t \tag{8.45}$$

8.5. CALCULATING TRANSPORT RATES AND PARTICLE DOSAGES

We can start by examining the penetration depth desired. Recall the traveling precipitation front model [Equation (8.45)]:

$$P = \frac{\left(\dfrac{Q_{eo} + Q_p}{\phi} + \dfrac{Q_{ep}}{f_v} \right)}{A \cdot N_D} \cdot t$$

Assume electroosmosis and hydraulic flows are negligible. Now we have:

$$P = \frac{Q_{ep}}{f_v \cdot A \cdot N_D} \cdot t$$

Recall Equation (8.38):

$$v_{ep} = \frac{Q_{ep}}{f_v \cdot A} \mu_{ep} \cdot E$$

Inserting yields

$$P = \frac{\mu_{ep} \cdot E}{N_D} \cdot t$$

This gives us a simple penetration expression for a simple structure, in which hydraulic seepage is not an issue and electroosmosis is considered negligible.

We can now turn our attention to the mobility of the nanoparticles. In general, all nanoparticles exhibit a mobility within an order of magnitude of each other, typically in the range of 10^{-7} to 10^{-8} m/s. We will examine how we calculate the amount of time needed for nanoparticles to arrive at steel reinforcement in order to provide corrosion protection.

We shall consider steel reinforcement that is covered by 0.3 m of concrete. The amount of time necessary to penetrate this depth of concrete must be determined. Solving the previous expression for t and inserting values recently discussed yields the following.

$$t = \frac{N_D \cdot P}{\mu_{ep} \cdot E} = \frac{0.3 \text{ m} \cdot (6)}{\left(\dfrac{10^{-8} \text{m}}{\text{volt} - \text{s}}\right) \cdot \left(\dfrac{40 \text{ volt}}{0.3 \text{ m}}\right)} \cdot \left(\frac{1 \text{ hr}}{3600 \text{ s}}\right) \cdot \left(\frac{1 \text{ Day}}{24 \text{ hr}}\right) = 15.6 \text{ Days}$$

A voltage of 40 V was selected, because it is the upper limit currently used in electrochemical chloride removal applications. We used an N_D value of 6 degrees of freedom. As noted earlier, this parameter effectively increases the assumed length of travel. It is assumed here that the selected voltage does not generate an electric current exceeding 1 A per square meter.

In this case, exceeding 1 A per square meter would not be expected to cause a problem, because the nanoparticles are expected to occupy all the volume of concrete subjected to the current. With nanoparticles present, the electric current damage is being compensated. This "protection" against higher current has been found to occur with current densities as high as 10 A per square meter.

8.6. PARTICLE SIZE AND DOSAGE LEVELS

As noted in Section 4.2 on crack repair, when particles pack into pores, the resultant reduction in porosity ranges from 30 to 60%, with 50% being a very typical result. It is interesting to note that the size of the particle does not seem to have a large influence on porosity reduction. This is a bit counterintuitive, since smaller particles can be packed into smaller pores; but in fact packed areas will exhibit an increasing amount of interparticle empty space as the particles get smaller, and this works against the benefit of smaller particles packing into smaller pores.

Though particle sizes must be small to enter pores, they need not enter all the smallest pores in order to provide benefits. Pores in functional concrete range from sub-nanometer up to 1 mm in size (for entrained air). In general, pores that influence strength and permeability run from 10 nm to 1 mm in size (Aligizaki, 2006). Since they will exhibit broad and unpredictable changes in diameter, the particles targeting them need to be smaller in size than the pores. In the author's prior work, a 20-nm silica particle has produced useful strength increases in young concrete with a 0.5 water/cement ratio (Cardenas and Goli, 2006). Polymer particles of 60 nm have also exhibited strength increases for such a concrete mix. Similar benefits were also observed when 2-nm particles of alumina were used.

On the basis of prior work, a 50% porosity reduction is not a bad assumption for work in concrete. This result can be enhanced under specific circumstances; but, in general, if particles are piling up against an embedded electrode such as rebar, a 50% porosity expectation is reasonable (Kanno *et al.* 2009). If particles are being driven through a section without the possibility of building up at an electrode, then something in the 30% range would be more likely.

If the treatment lasts several weeks, porosity reductions exceeding 70% are possible, because the electric field has the time to deposit a significant amount of other ions that are part of the system. In concrete, these may be ions liberated from the microstructure, such as sodium, potassium, or calcium. In the particle suspension fluid there may be other stabilizing ions that deposit within the pores, and more specifically within the interparticle spacing. If these other species are reactive with the nanoparticles or other phases, the material that actually provides porosity reduction may be further densified.

In Figure 8.5, particles are driven out of a cellulous sponge into the pores of concrete. Beneath the concrete, a metal conductor completes delivery of the current. Steel reinforcement can also be used to complete the circuit. In any case the idea is to pack nanoparticles into pores. Above the concrete, in the sponge, particles are not in contact with each other, and the volume they occupy in suspension will differ from the volume of pores they will occupy after treatment. If a 50% porosity reduction is anticipated, then we also anticipate that the volume of pores occupied by these particles will be twice the volume that the particles occupy while they are still in the sponge. This relationship is very helpful to remember when designing the dosage for treatment.

For a suspension containing 2 liters of particles, a 50% porosity reduction means that the particles will be packed into 2L/0.5 = 4 liters of pores. In some cases the porosity reduction will not go as well, perhaps due to pore size or tortuosity in the concrete. In a case like this, the porosity reduction may only be 30%. This means that 2 liters of particles will not pack the pore structure as efficiently, they will not be able to access every section of a given porous path, there will be gaps where no particles are present. In such a case, the calculation tells us that the particles will occupy (however inefficiently) 2 L/0.3 = 6.6 liters or pores. Clearly every section of pore in this sample is actually being occupied by particles. This calculation simply indicates that the treatment will spread these particles over a pore structure that is 6.6 liters in volume—which can be confusing, because the calculation gives the impression

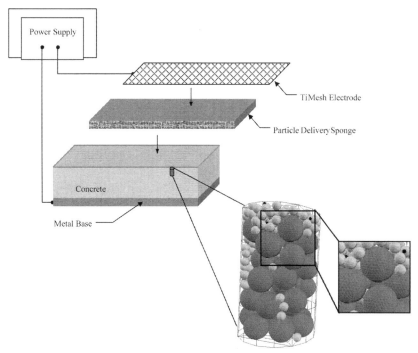

FIGURE 8.5. For a 50% porosity reduction, the volume of pores that the particles will oc-cupy following treatment will be twice the volume that they occupied while still suspended in the treatment fluid.

that the particles have expanded in volume after entering the pores. In reality, the treatment was simply not able to pack them more densely.

We now turn our attention to calculating the amount of suspension fluid needed to provide this porosity reduction. The first consideration is the volume of concrete treated. In the case illustrated in Figure 8.5, the concrete occupies 8 liters of space. If the porosity of this concrete is 20%, then the volume of pores available to treat is $8 \times 0.2 = 1.6$ liters. If we anticipate a 50% porosity reduction, the volume of particles needed to pack these pores is $1.6 \times 0.5 = 0.8$ liters. The sponge above the concrete has a volume capacity of only 1 liter. If the suspension liquid only holds 10 volume percent of particles, this means that the sponge only has 0.1 liters available for treating the concrete. Thus the suspension liquid will need to be replenished with a fresh batch 7 times after the initial load of particles is loaded into the concrete. Alternatively, we can provide a continuous re-circulation flow of particle suspension fluid.

8.7. ESTABLISHING AND MAINTAINING PARTICLE DELIVERY CIRCUITS

In order for nanoparticles to gain access to concrete pores there must be a reliable means of maintaining wet electrical contact between the particle suspension and the concrete pore fluid. Four main concerns come into play: suspension fluid handling, moisture in the concrete pores, the generation of evolved gases, and corrosion at contact surfaces. The following sections cover the practical aspects of these considerations. More specific examples of these approaches are available in the literature (Kupwade-Patil *et al.* 2011, Cardenas *et al.* 2011, and Cardenas *et al.* 2007).

8.7.1. Suspension Fluid Handling

A given treatment requires contact between the nanoparticle suspension and the surface being treated. Requirements for maintaining this contact vary, depending on the orientation and geometry of the structure being treated. The following sections describe how nanoparticle fluid contact with the structure is maintained for the principal cases that arise.

8.7.1.1. *Treating Horizontal Surfaces*

The easiest way to place a fluid suspension in contact with concrete is to pond it on top of the surface. Figure 8.6 shows a plexi-glass pond glued to the top of a concrete slab. This works well for bridge decks, floor joints, and other horizontal applications. On larger surfaces sand bags may be used as well. In another variation, a pre-cast section of concrete may be immersed in a particle fluid suspension. In all cases the application is relatively easy to maintain. Controlling leaks is one issue. Another is making sure that evaporation does not leave the pond electrode out of contact with the pond fluid.

8.7.1.2. *Treating Vertical Surfaces*

For vertical surfaces ponding is not an option, and the treatment process is a bit more dynamic. One way is to use a re-circulating sponge electrode assembly as shown in Figure 8.7. The particle suspension is pumped to a perforated tube that disperses the particle fluid

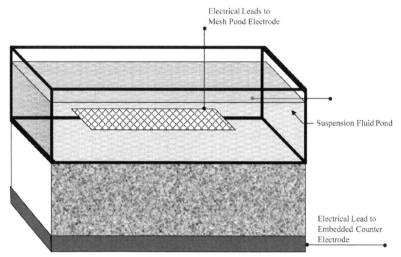

FIGURE 8.6. *Horizontal nanoparticle delivery setup. Ponded suspension fluid is used to drive nanoparticles into concrete pores using electrodes inside the pond and under, or within, the concrete.*

across the top of the sponge. The fluid flows through the sponge electrode, ensuring the wet electrical contact between the metallic mesh in the sponge and the capillary pore fluid in the concrete. The back board of the electrode, made of wood or plastic, helps contain the particle fluid and facilitates handling and placement of the assembly. At the bottom side of the sponge, the particle suspension is captured and recirculated. The fluid in the capture chamber is refreshed or changed as needed.

During a given treatment, the two flows Q_{ep} and Q_P represent the electrophoretic volume flow of particles into the concrete, and the volume flow rate of the particle suspension as it passes down the wall. The rectangular control volume surrounding the Q_{ep} term in Figure 8.7 is thus characterized by the volume flow of particles into the concrete and the volume flow rate of fluid through the sponge. In an ideal operation, the volume of nanoparticles inside the control volume drops to zero as the control volume completes its transit along the wall. In actual application, it is at least sufficient to ensure that this control volume is not empty of nanoparticles before it reaches the bottom. In this type of setup it is a simple matter to dose the system with a volume of nanoparticles sufficient to meet the porosity reduction requirements of the concrete, and simply permit the fluid to recirculate until the nanoparticle content in the reservoir has been depleted.

8.7.1.3. Treating Sloped Surfaces

The setup for sloped surfaces is intermediate between that for horizontal and vertical surfaces, and shares characteristics of both. At the same time, it is easier to maintain than either of the other cases. Figure 8.8 shows a 7 ft beam wrapped in a sponge electrode on 3 sides. The

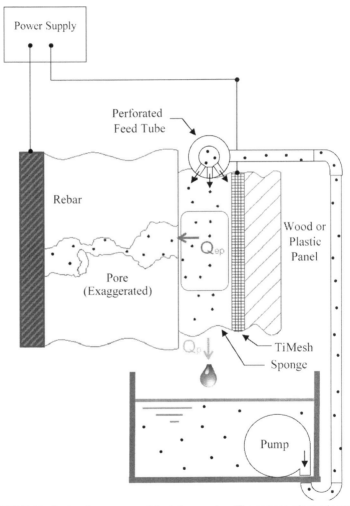

FIGURE 8.7. Vertical surface nanoparticle delivery setup. The sponge electrode receives a flow of particle suspension fluid released from a perforated hose at the top. As the fluid passes along the wall, the voltage strips particles out of the flow and into the concrete pores. At the bottom of the wall, the fluid is captured and recirculated back to the top of the sponge.

FIGURE 8.8. Sloped surface nanoparticle delivery setup. A recirculating sponge electrode applies treatment to a 7-ft reinforced concrete beam. The cellulose sponge is covered by diamond shaped Ti metal mesh as well as 1/4-inch plexi-glass panels to facilitate treatment flow observation. The particle suspension was fed onto the top of the sponge at the highest elevation point (left end of beam). The particle suspension flowed along the sponge while the Ti mesh drove the nanoparticles directly to the steel reinforcement within the beam. Image reproduced from Kupwade-Patil, K. (2010). Mitigation of chloride and sulfate based corrosion in reinforced concrete via electrokinetic nanoparticle treatment, Ph.D. Thesis, Louisiana Tech University, Ruston, LA.

negative pole of the power supply is connected to the steel reinforcement of the beam. The positive pole is connected to the titanium mesh within the sponge electrode. In this corrosion treatment, chlorides are extracted from concrete while positively coated pozzolanic particles are driven to the reinforcement. The sponge and mesh are also covered by acrylic panels in order to help control the particle suspension flow and facilitate observation. Particle suspension is fed to the sponge at the high point of the sloped beam, which is at the left end in Figure 8.8. The fluid flows down stream toward the right end of the beam, where it spills off into a catch basin containing a submersible pump. The pump drives the particle suspension from the catch basin back up to the release point at the high end of the beam.

The sloped surface is easier to maintain since there is less necessity to control leaks. During treatment leaks must always be controlled sufficiently to keep the sponge wet. It is not critical to stop leaks as long as their runoff is collected and incorporated into the recirculating fluid. In Figure 8.8, a long sheet of plastic is used under the beam to catch leaking fluid and guide it to the recirculation reservoir located at the right side of the beam. The submerged pump at this location returns the fluid to the feed tube located at the left side.

8.7.2. Gas Evolution

Electrodes connected to the negative pole of the power supply may generate hydrogen gas. As hydrogen is explosive, measures are required to avoid its collection. If there are chlorides in the concrete, this electrode will also produce chlorine gas, which is a health issue for the occupants and a corrosion problem for the instruments. It must be properly vented to the atmosphere.

Under certain circumstances, gas evolution can slow or stop particle transport. If the particle carrier suspension exhibits a pH well below 7, reactivity within the basic conditions in the concrete may be a concern. When acids are used to etch concrete, the phases containing calcium may decompose, producing gas. If gas forms at the concrete surface, it can also form bubbles within the pores—presenting a significant obstacle to particle transport. In such cases, the bubbles appear to dissipate after the first 3–5 days of treatment. This phenomenon is not well understood; but it appears to resolve itself after a few days.

8.7.3. Water Content of Concrete Pores

The electrokinetic processes that drive nanoparticles depend upon liquid paths for transport. The primary path is the network of capillary pores in hardened cement. For optimal transport, these pores need to be saturated with water. If the treatment is conducted on relatively fresh concrete, it is important to maintain a moist cure, which must be maintained at least to the point where treatment begins. Once the concrete is allowed to air-dry, the capillary tension developing inside the pores causes shrinkage. This shrinkage, and other desiccation impacts, will cause pores and volume porosity to reduce. In addition, the continuity between some of the pores will disrupt. These influences work against the goal of good treatment penetration throughout the material. The majority of such desiccation changes cannot be recovered after the concrete is rewetted. It is best, therefore, if relatively fresh concrete be kept wet up to the time when the electrokinetic treatment begins.

For older concrete moisture content is also an issue. In these cases, re-wetting the concrete is an important objective. One option is to simply start applying treatment with the expectation that the pores will become re-saturated through the natural process of capillary draw. Because of this phenomenon, re-wetting can be effective within hours. This is true especially for exterior concrete, because these materials are often only desiccated to about an inch in depth. No pressure need be applied, simply providing a spray of water to a structure is sufficient. Once capillary draw has pulled the water into about an inch in depth, electrical continuity throughout the material should be sufficient for an effective treatment.

8.7.4. Corrosion of the Circuit Contacts

Maintaining the treatment circuit for several days requires consideration of possible corrosion issues. The electrical contact between immersed or embedded electrodes and the circuit wiring can be problematic. Dissimilar metals are often in electrical contact, and if the points of contact between them are immersed in suspension fluid, rapid corrosion will soon disrupt the treatment circuit. Also, if a small leak emanates from the treatment pond or sponge and makes contact with an electrical connection point, corrosion will soon compromise the circuit at this point as well.

In any case, even if corrosion does not cause a radical disruption of the circuit, it can seriously slow down the delivery of particles, because particle transport is dependent upon the voltage drop between the two treatment electrodes. Often a drop in current means that particles are penetrating pores; but if corrosion begins at any point of contact, a significant voltage drop may quietly develop. This effectively reduces the amount of voltage drop available to actually drive the particles. The dropping current in this case becomes a false-positive indication of an effective treatment.

8.7.5. Circuit Polarization

As just noted, a drop in current may be a useful indicator of treatment progress; but the relationship between current drop and the relative success of the electrokinetic treatment is not well understood. In one respect the drop in current is clearly related to the reduction of available charge carriers. As treatment progresses, more of these charge carriers will lose the capacity to carry charge. Some will become lodged in pores. Others will react with either the solid or liquid phases present in the pore, and then fall out of suspension. In either case, the particle can no longer sustain electrokinetic transport. With fewer charge carriers available to sustain the electric current, this constitutes an increase in circuit resistance, and the current drops.

Other factors also cause current drops. The most common of these is charge polarization. In this case, the ions present in the fluid drift toward the pole bearing the opposite charge. The gathering of these charges will cause a decrease in current. One way to see if current drop relates to the success of particle lodging is to note the change in drop magnitude when nanoparticles are not present. Such comparisons can be made when nanoparticles are directed into pores that oppose them with pressure-driven fluid flow. When the flow is too high, the nanoparticles are unable to drift upstream. In this instance, current drop is not associated with particles being lodged into the pores.

Table 8.1 displays current drop results for an experiment conducted with cylinders of hardened cement paste that experienced pressure flow.

In each case the nanoparticles were driven upstream to see if they would lodge in the pores. The same electric field was applied in each case. In every case where the water/cement (w/c) ratio was 0.8, the flow was relatively slow and the nanoparticles penetrated the pores and reduced the coefficient of permeability by at least a factor of 10. The corresponding current drop ranged from 17–32%. In cases where the w/c ratio was 1.0, the fluid flow through these specimens was higher, and the nanoparticles were not able to enter the pores. As a result, none of the treatments produced a significant reduction in the coefficient of permeability. Since the parameter is logarithmic in nature, a drop in this coefficient needs to be no less than a factor of 10 before it can be considered significant. In general, it was observed in cases with an impact factor of at least 10, that the treatment current had dropped by at least 17%. In this way the relative amount of current drop indicated whether a given nanoparticle treatment achieved a successful amount of penetration.

Polarization often involves a sharp drop in current followed by a relatively slow descent. Figure 8.9 shows an electric current plot of a high-alumina cement concrete structure that was treated for a penetration of 3/4 inch down to the steel reinforcement. It is a typical example

TABLE 8.1. Current Drop as an Indication of Nanoparticle Treatment Success on Hardened Cement Paste Subjected to an Opposing Hydraulic Flow.

Treatment	Paste w/c Ratio	Paste Alkali Content	Permeability Impact Factor	Current Drop (%)
Colloidal alumina (2 nm)	0.8	Low	81	37
	0.8	High	17	17
	1.0	Low	1	17
	1.0	High	2	13
Colloidal silica (20 nm)	0.8	Low	186	32
	1.0	Low	1	17
	1.0	High	37	24
Sodium silicate (range)	0.8	Low	26	32
	1.0	Low	1	11
	1.0	High	1	25

drawn from unpublished work conducted at the Applied Electrokinetics Laboratory at Louisiana Tech University in 2011 with extensive assistance from graduate student Mir Al-Masud. During the 14-day treatment period, the current was initially over 32 mA per square foot of concrete surface area. The current leveled out at approximately 6 mA/ft^2 and stayed at this level until day 9. In reality, the current in this steady region of the plot still continued to drop, but much more slowly. The current dropped a bit more markedly at day 10 and 11. It is not clear why this occurred. These anomalies are also common in nanoparticle treatments.

In this case the current drop was over 80% in one day. Changes to the material were also significant. The treatment utilized a 24-nm alumina-coated silica particle that carried a positive charge. In response to the treatment, porosity of this high-strength concrete dropped by 49%.

It is interesting to note that any change in the treatment process causes the current level to change. Even the simple act of shutting off the circuit and turning it back on may cause ions to drift away from the poles, causing the current to climb. In the present example, nanoparticle treatment was stopped and a solution of calcium sulfate was used to drive in sulfate ions. After several hours the polarity of the circuit was reversed in order to remove sulfate ions while calcium ions were being injected. Prior to being fully removed, the remaining sulfates were accelerating reactions between the calcium and any alumina or silica particles that were previously injected. The sulfate ion can help catalyze the forma-

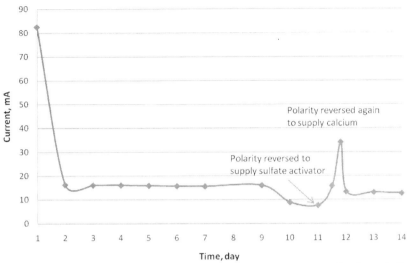

FIGURE 8.9. Plot of current delivered during 14-day particle treatment. The first polarity reversal was conducted when sulfate ions were injected into the concrete for a period of 6 hours. The second polarity reversal was conducted to extract sulfates while delivering calcium ions as a surface-hardening treatment that was sustained for the remainder of the treatment period.

tion of additional cementitious materials, but it must be removed using reverse polarity so that it does not cause sulfate attack. During the sulfate treatment, current spiked sharply before it was reversed again and dropped back down to the 12 mA (5 mA/ft^2) level.

8.7.6. Electric Field Distribution Control

Depending upon the shape and spacing of the electrodes used to apply a driving voltage, nanoparticles may be limited in their ability to permeate the entire concrete structure. Figure 8.10 compares the electric field lines associated with a pair of flat electrodes and a pair point-source electrodes. In the flat pair, the electric field lines are straight and parallel. Each nanoparticle traveling in this electric field has the same distance to travel through the specimen, and each path exhibits the same potential drop. For the point-source electrodes, the electric field lines curve through space. The central lines constitute the path of least resistance for the nanoparticles—meaning more nanoparticles are driven along this route, resulting in uneven distribution. It is possible that some portions of the concrete structure will receive no nanoparticles, at least initially.

As treatment progresses, the number of particles traveling the path directly between the two electrodes will diminish. As this region fills with nanoparticles, it will no longer constitute the path of least resistance. More of the nanoparticles will then follow the outer field line paths, as they in turn become new paths of least resistance. Because material property changes may not be uniform throughout the specimen, reductions in porosity or other changes may take longer to develop throughout the structure.

8.7.7. Electrode Design and Placement

In many instances the choice of electrode material depends on where it is located in the treatment circuit. The most important question to answer is which pole of the power supply will be connected to the electrode. In general, almost any metal can be connected to the negative pole of the power supply without concern for degradation. Much more careful material selection is required when attaching a metal electrode to the positive pole. Most metals will corrode rapidly under these circumstances. The results include staining from the dissolving electrode, loss of electrical continuity in the treatment circuit, and possible loss of structural integrity, due to concrete cracking and degradation of the reinforcement.

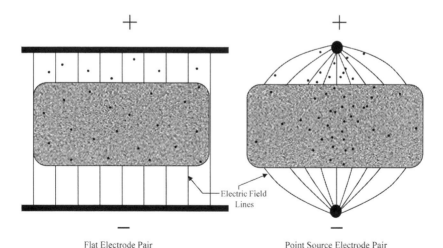

Flat Electrode Pair   Point Source Electrode Pair

FIGURE 8.10. The shape of the electrode can influence how nanoparticles are distributed throughout a given concrete structure. The beam in the uniform electric field receives a more uniform distribution of nanoparticles during treatment.

Materials used for impressed current cathodic protection systems tolerate connection to the positive pole. A common example is mixed-metal oxide-coated titanium. It is generally available as an expanded metal mesh and in wire form. This ceramic-coated metal is durable enough for water desalination and the production of chlorine gas.

The next consideration is the shape and orientation of the structure being treated. As discussed earlier, the two primary orientations are horizontal and vertical delivery. Another key consideration is the available access to a given surface. Treatment surface access is a challenge that may involve contact with soil, water, or an interior structure. The following sections illustrate approaches that can be taken to address these access and orientation modes.

8.7.7.1. Electrodes for the Bridge Deck

The bridge deck is typically an 8-inch thick section with two layers of steel reinforcement. Treatments are designed to draw out chlorides while nanoparticles are layered around the top layer of steel. Figure 8.11 illustrates the setup used to accomplish it. A pond of particle fluid is established at the top surface. The positive pole of the power supply is attached to the mixed-metal oxide-coated titanium mesh (Ti). The negative pole of the power supply is connected to the top layer of the deck reinforcement. In general, the electrical continuity of this reinforcement network is good; but it is prudent to verify this by boring down to it at several locations and using an ohm meter to ascertain low resistances. The leads used to connect to the reinforcement grid need to be coated down to and including the point of the connection, so that

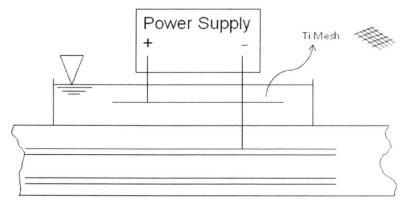

FIGURE 8.11. Electrokinetic nanoparticle treatment setup for a bridge deck.

galvanic corrosion does not cause a loss of continuity during treatment. This coating also ensures that the electric field delivers nanoparticles to the reinforcement, not to various locations along the connecting wire. A DC current is applied using 40 V DC. Higher voltage is avoided from a worker safety stand point. Historically, care is taken to ensure that the current density is at or below 0.1 A per square foot of concrete surface. Nanoparticle delivery allows for much higher current densities. The damage associated with high current densities does not materializing when nanoparticles are delivered. This nanoparticle strength compensation has been found to work at current densities that are 5–10 times higher that 0.1 A/ft^2.

At low current densities a typical chloride extraction process is in operation for 6–8 weeks. The use of nanoparticles and the higher processing currents that they enable for these applications should bring this time requirement down significantly, to perhaps 2–3 weeks.

8.7.7.2. Electrodes for Columns and Walls

Columns and walls require the use of a recirculating sponge electrode assembly, such as that illustrated earlier in Figure 8.7. For a column, the sponge electrode can be fashioned as a wrap (Figure 8.12). As with the bridge deck, the positive pole of the power supply is attached to the Ti mesh inside the sponge electrode. The negative pole of the circuit is attached to the reinforcement within the column. The plastic outer layer of the electrode may consist of shrink wrap. For extremely tall columns, the head pressure could be alleviated by establishing several shorter sub-sections wrapped for treatment and equipped with separate particle suspension feeds. Each of these subsections could still share the same power supply, as long as there is sufficient current capacity.

For wall sections that have steel reinforcement, the setup is very similar to the column method, except that the sponge electrode does not need to be wrapped. If there is no reinforcement in the wall, then a sponge electrode on both sides of it works well.

If the column section or wall needs complete particle treatment but the structure contains steel reinforcement, then special attention to corrosion is warranted. If the reinforcement is not wired into the treatment circuit, it will be susceptible to stray current corrosion. In this type of corrosion, the side of the reinforcement nearest to the negative pole of the circuit will start dissolving. To prevent or minimize this type of damage, an additional power supply must be used to cathodically

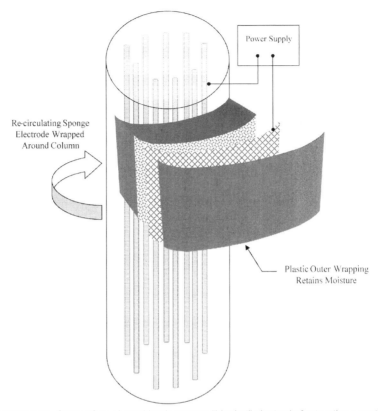

FIGURE 8.12. *Setup of an electrokinetic sponge "blanket" electrode for treating a vertical column.*

protect the reinforcement while the treatment is underway. Ideally, this power supply should be part of a potentiostatic device that monitors corrosion potential of the reinforcement and automatically provides enough protective current to prevent stray current corrosion.

8.7.7.3. Electrodes for Floors and Foundations

In floors and foundations, the soil-side of the structure makes access a little more difficult to achieve. The setup resembles a horizontal treatment, except that the soil-side electrodes are installed deeply enough to reach the soil-side of the foundation. As illustrated in Figure 8.13, the outside surfaces of these soil-side electrodes must be electrically isolating over all portions except the ends. The top end of the electrode receives a feed of nanoparticles as well as the electrical lead to the power supply. As long as it is connected to the negative pole of the power supply, this tube will remain protected against corrosion. At the bottom end of the floor electrode, the tube is also uncoated and perforated, so that particles can exit into the soil and then travel up into the concrete slab. This configuration provides a means of conducting sulfate extraction followed by strength recovery and sealing of the bottom surface to mitigate the re-entry of sulfates.

If simple hardening or permeability reduction of the top surface is required, then a simpler setup could be used. In such cases the particle suspension could be ponded at the surface. If available, reinforcement could be used as the counter electrode. Here again, some care is required to avoid accelerated corrosion by using a positively charged particle that permits the connection of the reinforcement to the negative pole of the power supply.

If no reinforcement is present, an attempt must be made to use coun-

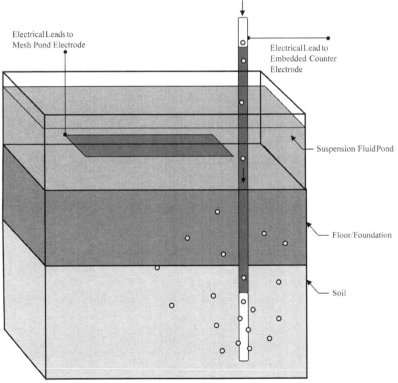

FIGURE 8.13. Nanoparticle treatment setup for recovery from sulfate attack on a structural foundation or floor slab.

ter-electrodes driven into the soil at locations surrounding the concrete slab. This creates a much longer treatment circuit, which naturally produces a lot of resistance. The setup requires driving nanoparticles from above the slab. The long circuit path means that nanoparticles near the slab perimeter are driven preferentially, while those further away make relatively little progress. This could be managed by simply treating the peripheral regions first, then working toward more distant locations until the closer sections experienced significant increases in resistance (both to current and nanoparticle flow). If the soil beneath the slab is too dry to enable a reasonable current draw (> 0.01 A/ft^2), then the use of the floor-penetrating electrode of Figure 8.13 is indicated. The distributed network of these electrodes is used to get the soil sufficiently wet to facilitate particle transport.

8.7.8. Steel Reinforcement and the Positive Pole

Sometimes the only way to proceed is by connecting the reinforcement to the positive pole. This should not be attempted in locations were structural failure is a possible outcome. Strictly for non-structural cases, there is a way to proceed, which involves careful preparation of the reinforcement surface. A coating must be electrokinetically applied and later sacrificed during treatment. The following sections describe unpublished work that would make this type of treatment possible for a medical procedure that is under development at the Applied Electrokinetics Laboratory at Louisiana Tech University. The data presented was acquired with extensive assistance from Kunal Kupwade-Patil, Anupam Joshi, and Sahaja Patel, graduate students in construction materials engineering, nanosystems engineering, and biomedical engineering in 2009.

The aim of this work was the development of a counter-electrode which would enable the electrokinetic transport of calcium-rich particles for the reduction of porosity in osteoporotic bone tissue. It was important to obtain a treatment process that did not permit the dissolution of dangerous metallic ions into the patient's tissues. The idea was to coat the steel electrodes with oxides of metals that are non-toxic, such as sodium, potassium, and calcium. Several combinations of metal hydroxides were tested to see which ones produced electrodeposits that were both sacrificial and also protective. They needed to be re-dissolvable when the electric current was flowing, but they could not permit significant dissolution of the underlying iron.

Steel specimens were subjected to hydroxide electrodeposition at

0.1 A/ft^2 for 24 hours. The hydroxides of sodium and potassium were provided at a 1 molar concentration. Calcium hydroxide, due to its low solubility, was dosed at only 25 mmole/l. The appearance of these specimens following electrodeposition is shown in Figure 8.14. All the specimens except for controls showed relatively little change in color. The corrosion potential data obtained immediately after electrodeposition are listed in Table 8.2. In all cases the corrosion potentials were strongly negative, indicating an active coating. The most negative case was that of the controls, which possessed no protective coating.

After 6 hours of simulated treatment, the corrosion potentials were measured again but this time in simulated body fluid. These values are listed in Table 8.3. In every case the Vcorr values were found to shift negatively. The most negative value, of –2.169, was exhibited by the

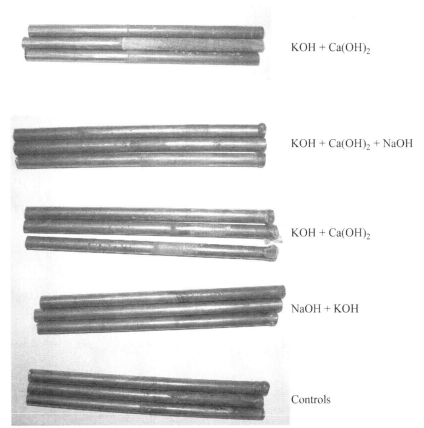

KOH + Ca(OH)$_2$

KOH + Ca(OH)$_2$ + NaOH

KOH + Ca(OH)$_2$

NaOH + KOH

Controls

FIGURE 8.14. 1020 Steel after electrodeposition treatment applied at 0.1 A/ft^2 for 24 hours.

TABLE 8.2. Corrosion Potentials of Electro-coated Steel After Deposition Treatment for 24 Hours at 1 A/m².

Electrodeposition Fluid	Vcorr* Trail 1 (Volts)	Vcorr* Trail 2 (Volts)	Vcorr* Trail 3 (Volts)	Vcorr* Average (Volts)
KOH +Ca(OH)$_2$ + NaOH	−0.984	−1.467	−0.868	−0.866
KOH + Ca(OH)$_2$	−0.898	−0.859	−0.978	−0.878
NaOH + KOH	−0.978	−0.467	−0.759	−0.786
NaOH + Ca(OH)$_2$	−0.676	−0.841	−0.858	−0.791
Control	−0.787	−1.578	−0.988	−0.889

*Vcorr measured in respective electrodeposition fluids.

combination of NaOH + Ca(OH)$_2$. This appears to correlate to large deposits later observed for this case (Figure 8.15).

Figure 8.15 shows steel specimens following electrodeposition and simulated treatment. The specimens electro-coated with all three hydroxides exhibited the cleanest appearance. The worst case was the steel coated with sodium and calcium hydroxide. This was also the case that exhibited the most negative corrosion potential.

Atomic absorption data for iron content among these cases are shown in Table 8.4. Here again, the case involving all three hydroxides performed the best by releasing the least amount of iron into the treatment SBF used for the simulated treatment. The concentration of iron released in this case was 1.79 ppm. The worst case, with a release concentration of 11.79 ppm, was the sodium and potassium hydroxide coating. Interestingly, this case did not exhibit the worst appearance in the comparisons presented in Figure 8.15. The worst appearance in Figure 8.15 correlated to the second worst concentration of 7.69 ppm.

TABLE 8.3. Corrosion Potentials of Electro-coated Steel After Simulated Treatment for 6 Hours at 0.1 A/ft².

Electrodeposition Fluid	Vcorr* Trail 1 (Volts)	Vcorr* Trail 2 (Volts)	Vcorr* Trail 3 (Volts)	Vcorr* Average (Volts)
KOH +Ca(OH)$_2$ + NaOH	−1.250	−0.985	−1.102	−1.112
KOH + Ca(OH)$_2$	−1.998	−1.064	−0.806	−1.289
NaOH + KOH	−0.877	−1.477	−1.396	−1.241
NaOH + Ca(OH)$_2$	−1.985	−2.156	−2.366	−2.169
Control	−2.133	−1.663	−1.261	−1.685

*Vcorr measured in SBF.

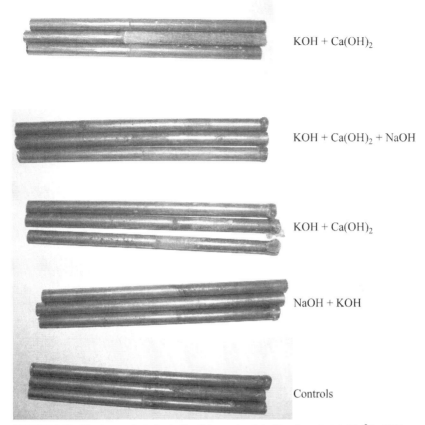

KOH + Ca(OH)$_2$

KOH + Ca(OH)$_2$ + NaOH

KOH + Ca(OH)$_2$

NaOH + KOH

Controls

FIGURE 8.15. *Electro-coated steel after 6-hour simulated treatment at 1 A/m^2 in SBF.*

TABLE 8.4. Atomic Absorption Results for Iron Released After 6-Hour Simulated Treatment in SBF at 0.1 A/ft^2.

Type of Exposure	Average (ppm)
KOH +Ca(OH)$_2$ + NaOH	1.79
KOH + Ca(OH)$_2$	4.44
NaOH + KOH	11.79
Ca(OH)$_2$ + NaOH	7.69
Control	3.28

Each result is an average of 3 trials.

141

This work demonstrated that a nanoparticle transport process could be staged using simple iron electrodes connected to the positive pole of the circuit without causing significant degradation of the steel. The electro-deposited coating was able to function as a sacrificial coating that protected the steel against rapid dissolution.

8.8. EXAMPLE NANOPARTICLE TREATMENT DESIGN: CONCRETE BRIDGE DECK

In this design example, the circumstances of a real world application makes the transport equations simpler. In contrast, there are still some interesting subtleties regarding the dosage of nanoparticles that need careful attention. The following sections demonstrate how a given application can boil down to a fairly simple set of issues.

8.8.1. Design Example: Calculating the Treated Volume of Concrete

For obtaining a dosage level, the volume of deposited nanoparticles must be determined. This calculation requires knowledge of the amount and size of reinforcement, and the thickness of nanoparticle coating desired. Figure 8.16 shows the ring of protective particle deposit that is applied to the reinforcement.

Considerations:

*   1 m^2 of surface covers 10 m of reinforcement bar (surface layer only).

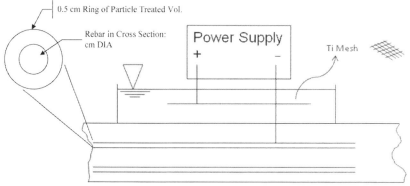

FIGURE 8.16. *Treatment design for nanoparticle corrosion projection of bridge deck.*

- 1 m^2 of concrete surface holds suspension pond of some depth.
- Must use positive particle (to prevent steel corrosion). Select: Alumina-coated Silica, 24 nm Dia.
- Apply 1/2 cm thick layer of particles around reinforcement.

Calculation:

$$\text{Treated vol.} = (10 \text{ m})(\pi)(r^2_{\text{treated}} - r^2_{\text{bar}})$$

$$= (10 \text{ m})(\pi)[(0.03 \text{ m})^2 - (0.02 \text{ m})^2] \cdot \left(\frac{100 \text{ cm}}{1 \text{ m}}\right)^3$$

$$= 6.3 \times 10^3 \text{ cm}^3 = \frac{6.3 \ell}{\text{m}^2 \text{ of deck}}$$

8.8.2. Design Example: Treatment Time Calculation

The penetration time tells us how long we can wait before we will need to add more nanoparticles to the suspension pond. The travel time through the pond is on average 6 times faster than through the concrete pore system. For this reason, we can neglect this distance for the time calculation. On occasion, we may use this distance in our calculation of the electric field strength. In this case we can assume the mesh is simply lying on the concrete surface, so this 5 cm distance (Figure 8.17) will not be included in the calculation. We use the simplified penetration model that ignores the impact of hydraulic flow and electroosmosis.

8.8.3. Design Example: Dosage Volume of Nanoparticle Suspension

Figure 8.18 illustrates the volume of particles that can occupy a given region of protection. Particles occupy more space after they consolidate in the pores.

8.8.4. Design Example: Treatment Specification Summary

In this design example, a setup was developed for treatment of a bridge deck structure. The nanoparticle dosage was calculated in terms of a unit amount of concrete surface area. A relatively slow particle mo-

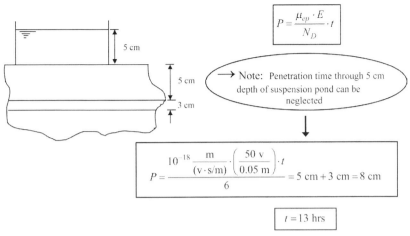

FIGURE 8.17. Treatment time calculation for bridge design example.

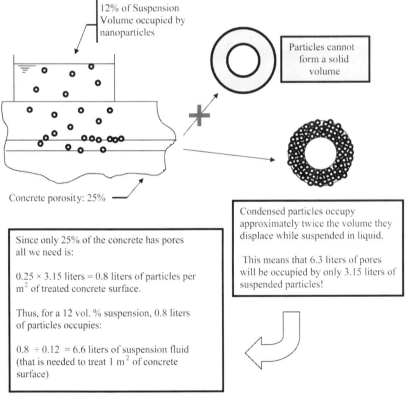

FIGURE 8.18. Calculation of volume of particle suspension needed to treat one square meter of concrete surface area that is above 10 m of 2-cm diameter reinforcement.

bility was assumed. This led to the calculation of a fairly conservative treatment time.

8.8.4.1. Layout

As shown in Figure 8.19, the treatment is ponded on the surface with the counter electrode lead attached to the steel reinforcement. The power supply is located in a weather safe enclosure.

8.8.4.2. Dosage

To treat 1 m^2 of concrete surface, 6.6 liters of alumina coated silica suspension are required. Distilled water may be added to increase the liquid height of the suspension as needed to completely submerge the drive electrode. The treatment will build up a 1 cm layer of particles on the 10 meters of reinforcement located below each square meter of concrete surface.

8.8.4.3. Treatment Time and Start Up

Treatment requires 13 hrs (~2 business days). This translates to as much as four business days when travel and setup are factored. When powering up, it is important to check that the current is not too high when applying 50V. Current engineering standards for chloride extraction and other activities call for a limit of 0.1 A/ft^2. The use of nanoparticles has enabled us to extend this limit to well over 0.5 A/ft^2.

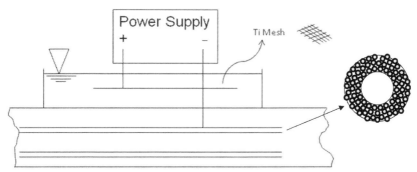

FIGURE 8.19. *Setup of nanoparticle treatment for bridge deck application.*

8.9. REFERENCES

Ashby, M. F., Jones, D. R., Engineering Materials, An Introduction to their Properties and Applications, *International Series on Materials Science and Technology,* Vol. 34, Pergamon Press, New York, p. 166, 1980.

Aligizaki, K. (2006) *Pore Structure of Cement Based Materials: Testing Interpretation and Requirements,* Taylor and Francis, London, UK.

Barbarulo, R., Marchand, J., Snyder, K., Prene, S., "Dimensional Analysis of Ionic Transport Problems in Hydrated Cement Systems Part 1. Theoretical Considerations", Cement and Concrete Research, Vol. 30, No. 12, pp. 1955–1960, December 2000.

Bockris, J. O. M. and Reddy, A. K. N. (1976) *Modern Electrochemistry,* Vol. 1 & 2, Plenum Press, New York.

Cardenas, H., Kupwade-Patil, K., Eklund, S., "Corrosion Mitigation in Mature Reinforced Concrete using Nanoscale Pozzolan Deposition", *American Society of Civil Engineers, Journal of Materials in Civil Engineering,* Vol. 23, No. 6, 13 June, 2011.

Cardenas, H., Struble, L., "Modeling of Permeability Reduction in HCP via Electrokinetic Nanoparticle Treatment," *American Society of Civil Engineers, Journal of Materials in Civil Engineering,* Vol. 20, No. 11, Nov. 2008.

Cardenas, H., Paturi, P., Dubasi, P., "Electrokinetic Treatment for Freezing and Thawing Damage Mitigation within Limestone", *Proceedings of Sustainable Construction Materials and Technologies, Coventry England,* 12 June 2007.

Cardenas, H. and Goli, N., "Investigation of Electrokinetic Nanoparticle Technology for Corrosion Mitigation in Reinforced Concrete", American Concrete Institute, *Proceedings of Concrete Solutions, St. Malo, Brittany, France,* June, 2006.

Carman, P. (1956) *Flow of Gases through Porous Media,* Academic Press, New York.

Chatterji, S., "Evidence of Variable Diffusivity of Ions in Saturated Cementitious Materials", *Cement and Concrete Research,* Vol. 29, pp. 595–598, 1999.

Cook, R., Hover, K., "Mercury Porosimetry of Hardened Cement Pastes", *Cement and Concrete Research,* Vol. 29, No. 6, pp. 933–943, June 1999.

Darcy, H. (1856) *Les fontaines publiques de la ville de Dijon,* Appendix: "Determination of the Law of the Flow of Water through Sand," Victor Dalmont, Paris, pp. 590-594.

Diamond, S., "Aspects of Concrete Porosity Revisited", *Cement and Concrete Research,* Special Issue, Vol. 29, No. 8, pp. 1181-1188, August, 2000

Garboczi, E. J. (1990) "Permeability, Diffusivity, and Microstructural Parameters: a Critical Review," *Cement and Concrete Research,* Vol. 20, No. 4, pp. 591–601.

Garboczi, E.J., Bentz, D., *Advances in Cementitious Materials,* American Ceramic Society, p. 365, Westerville, OH, 1991.

Glasstone, S. (1946) *Textbook of Physical Chemistry,* 6th Ed., D. Van Norstrand Company, Inc., New York.

Hayes, R., Drytronic Inc., private conversation, November. 1997.

Hearn, N., Hooton R. D. and Mills, R. H. (1994) "Pore Structure and Permeability," *STP 169C Significance of Tests and Properties of Concrete and Concrete-making Materials,* ASTM Publication Code number 04-16030-07, Chapter 25.

Hunter, R. J. (1992) *Foundations in Colloid Science,* Vol. 1, University Press, Belfast, Northern Ireland.

Kanno, J., Richardson, N., Philips, J., Mainardi, D., Cardenas, H., "Modeling and Simulation of Electromutagenic Processes for Multiscale Modification of Concrete", *J. of Systemics, Cybernetics and Informatics,* Vol. 7, No. 2, pp. 69–74, 2009.

Kupwade-Patil, K., Cardenas, H., Gordon, K., Lee, L., "Corrosion Mitigation in Reinforced Concrete Beams via Nanoparticle Treatment", *J. American Concrete Institute,* Submitted, August, 2011.

Marchand, J. *et al.* (1998) "Modeling Ionic Interaction Mechanisms in Cement-Based Materials—

An Overview," *Proceedings of the Sidney Diamond Symposium,* American Ceramic Society, Honolulu, Hawaii, September, pp. 143–159.

Mitchell, J. K. (1976) *Fundamentals of Soil Behavior,* Wiley & Sons, New York.

Morsy, M., "Effect of Temperature on Electrical Conductivity of Blended Cement Pastes", *Cement and Concrete Research,* Vol. 29, pp. 603–606, 1999.

Powers, T. C. (1960) "Physical Properties of Cement Paste," *Proceedings, 4th International Symposium on the Chemistry of Cement,* Washington, DC, Vol. 2, Paper V-1, pp. 577–613.

Reinhardt, H. W., "Transport of Chemicals through Concrete", Materials Science of Concrete III, J. P. Skalny, Ed., American Ceramic Society, Westerville, OH, pp. 209–241, 1992.

Samson, E., Marchand, J., Beaudoin, J., "Modeling of the Influence of Chemical Reactions on the Mechanisms of Ionic Transport in Porous Materials, An Overview", *Cement and Concrete Research,* Vol. 30, No. 12, pp. 1895–1902, December 2000.

Shaw, D. J. (1992) *Introduction to Colloid and Surface Chemistry,* 4th Ed., Reed Educational and Professional Publishing Ltd., Cornwall, England.

Stokes, R. H., and Robinson, R. A., *Electrolyte Solutions,* Butterworth's Publications, Ltd., London, 1955.

Taylor, H.F.W. (1997) *Cement Chemistry,* 2nd Edition, Thomas Telford Publishing, London, p. 249.

Zhang, T. *et al.* (2001) "Adsorptive Behavior of Surfactants on Surface of Portland Cement," *Cement and Concrete Research,* July, Vol. 31, No. 7, pp. 1009–1015.

CHAPTER 9

Electromutagenics

AT some point an electrokinetic treatment will cause enough structural and/or chemical change to the microstructure of concrete so that we can consider it a mutated version of the original design. These changes can be described in a hierarchical order. The lowest order is simple reduction in porosity. Next is the conversion of chemical phases achieved by an introduced agent. Then comes the development of entirely new phases obtained exclusively from introduced species. These changes can happen side by side with any changes occurring to the existing phases. The following sections describe examples of how such changes are achieved.

9.1. PORE STRUCTURE REVISION

When nanoparticles fill a pore without reacting with any other species, they can form a densely packed structure within the pore. The new porosity consists of interparticle spacing. The remaining porosity is located in pores that did not receive particles, as well as in locations at pore walls that are too small for an additional particle, but larger than the interparticle spacing.

In Figure 9.1 the results of mercury intrusion porosimetry tests are shown for several specimens of concrete, each with a 0.5 w/c content (Kupwade-Patil, 2010). The highest curve in this collection pertains to a control sample that did not receive a particle treatment. The curve shows that the pores ranged from 0.05 to 1000 microns. In addition, 25% of the pores were less than 10 microns. Next consider the case in

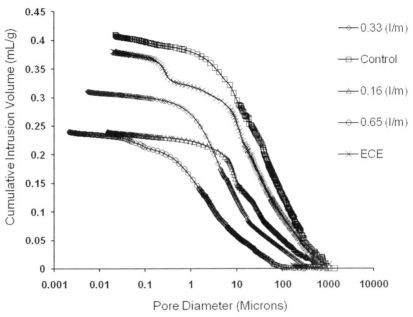

FIGURE 9.1. Mercury induced porosimetry (MIP) curves on powdered concrete two inches away from the concrete. Adapted from Kupwade-Patil, 2010 with permission.

which electrokinetic treatment was conducted with a 24 nm silica particle with a dosage of 0.65 l/m^2 of concrete surface. The total volume of pores in a gram sample dropped from 0.41 to 0.24 ml. The pore sizes ranged from 2 nm to 100 microns. Of these pores, 75% are less than 10 microns. The electrokinetic treatment clearly reduced the pore volume of the material. The sizes of the pores observed were also smaller.

Figure 7.2 illustrates the elimination of microcracks in limestone (Cardenas, *et al.*, 2007). Cracking was induced by 12 cycles of freezing and thawing at temperatures ranging from 0–70° F daily for 12 days. Cracking causes an increase in porosity. Cracks like the one observed in Figure 7.2 were no longer encountered after a treatment of calcium ions and sodium silicate. After treatment, the material was also relatively free of the extremely fine-grain detail.

The shrinkage and/or elimination of pores constitutes the simplest level of electromutagenic processing. In this case, the process was a little more complex, since it involved the introduction of a new phase (C–S–H) (Cardenas *et al.* 2007). It was achieved throughout a 5-inch wall section. The treatment was conducted using a vertically-oriented sponge electrode.

The porosity reduction in this case was approximately 40%. The change in strength associated with this case was 82%.

9.2. CHEMICAL ACTIVATION/MICROSTRUCTURAL PHASE REVISION

In this class of mutation, the electric field injects a species that reacts with phases present inside the solid matrix of the concrete. The new phase created may add to the quantity already present, or introduce a species that is closely related to those native to the system. The following sections describe how such transformations can be achieved.

9.2.1. Calcium Hydroxide Conversion

When silica and alumina nanoparticles are driven into pores of concrete, an opportunity arises for the conversion of calcium hydroxide into C–S–H and C–A–H phases. These are the pozzolanic reactions associated with silica fume, fly ash, and other sources of silica and alumina.

The fundamental pozzolanic reaction is the hydration of silica or alumina. In the presence of calcium hydroxide, these reactions may be expressed as follows: $CH + S + H \rightarrow C - S - H$ (calcium silicate hydrate), as well as the alumina analogue, $CH + A + H \rightarrow C - A - H$ (calcium aluminate hydrate), (Mindess *et al.* 2003, p. 42).

Initially, OH^- ions attack the O–Si or O–Al bonds of the mineral. As more of these bonds are broken, ions of silica or alumina enter solution, and the hydroxyl ions attach themselves to silicon or aluminum atoms. Silicate and aluminate anions develop and form an amorphous material balanced by H^+, K^+, and Na^+ ions from the capillary pore fluid. Later, available Ca^{+2} reacts with the amorphous material to form the final C–S–H or C–A–H type products. These reactions often proceed much more slowly and with lower heat evolution than is exhibited by the primary hydration reactions of Portland cement (Taylor, 1997, p. 280). Reactivity of flyash is also dependent on the structure of the individual particles. Flyashes with dense glassy surface layers can require pre-treatment with basic solutions in order to corrode surface layers and make interior sites accessible for enhanced reactivity (Fan *et al.*, 1999). In the case of silica fume, densified forms exhibit slower reaction rates, presumably due to the reduction in available bonding sites (Sanchez de Rojas, 1999).

Several variants of calcium aluminates have been observed in ordinary Portland cement as well as high alumina cements (Taylor, p. 298, 1990). Calcium aluminate can form chemically unstable phases, C_2AH_8 and CAH_{10}; and both of phases can transform to C_3AH_6. This new phase is higher in density that either of the others. This causes an increase in porosity. The use of alumina-coated silica nanoparticles causes a small amount of C_2AH_8 formation (Figure 9.2). It has the classic hexagonal crystal habit of such unstable phases. The new phase is cubic and significantly denser.

It is interesting to note that concretes found to exhibit C_2AH_8 retained higher strength than untreated companion specimens. Probably this is due to the alumina portion of a given nanoparticle constituting only 2% of the total particle mass, the rest being silica. Also, the insertion of particles causes porosity to drop and strength to increase even when the reaction is not taking place. In one case it was observed that only 8% of the calcium hydroxide in the concrete had been converted to C–A–H or C–S–H. Clearly the rest of the nanoparticles were not able to access the remaining calcium hydroxide in the concrete. The hardened cement paste phase can easily contain as much as 50% or more calcium hydroxide. Strength gain of the nanoparticle-treated concrete was probably influenced by C_2AH_8 only to a small extent. This means that the conversion C_3AH_6 is more than matched by the other strength-enhancing influences of the treatment.

FIGURE 9.2. Magnified image showing striated needles of di-calcium aluminate hydrate (C_2AH_8). Reproduced from Kupwade-Patil et al. June 2011 with permission from ASCE.

9.2.2. Monosulfate Conversion

Sulfates entering cement capillary pores come into contact with monosulfate. These species react to form ettringite as follows:

$$C_4 A\bar{S} H_{12} + 2Ca^{+2} + 2SO_4^{-2} \rightarrow C_6 A\bar{S}_3 H_{32} \qquad (9.1)$$

At some stage, gypsum may also form. Gypsum does not contribute to matrix destruction until sulfate content exceeds 1000 mg/l (Mindess *et al.* 2003, p. 487). Solid volume expansions due to the formation of gypsum and ettringite are reported at factors of 2.2 and 2.8 respectively (Zivica, 2000).

Long prismatic needles of ettringite as well as gypsum have long been believed to be the source of expansion leading to disruption of the cement matrix (Solberg and Hansen, 2001). More recent work has examined the role of C–S–H decalcification. This loss of calcium in the binder matrix has come to be recognized as a major source of strength reduction that could account for a good deal of concrete structure degradation (Taylor, 1997, p, 373). In other work (Odler and Colan-Subauste, 1999), a significant portion of expansion is associated with water imbibitions. Ettringite crystals of colloidal dimensions have been found to take up significant amounts of water osmotically, leading to expansion and damage (Mehta and Wang, 1982).

Differences in the severity of sulfate attack have been observed that depend on the counter ion. Especially severe degradative influence has been exhibited by $MgSO_4$ (Bonen, 1994; and Rasheeduzzafar *et al.*, 1994). It has been suggested that the Mg^{2+} participates in the degradation process. When Mg^{2+} is present, brucite and $M_3S_2H_2$ precipitates have been observed in addition to ettringite (Gollop and Taylor, 1992). Ettringite formation utilizes calcium, leading to decalcification of C–S–H.

Calcium carbonate has been found to contribute carbonate ions that substitute for sulfate ions in ettringite and monosulfate structures (Kakali *et al.*, 2000). Due to the relatively low solubility and high stability of calcium carbonate, it was found that a partial substitution of calcium carbonate for ordinary Portland cement delayed the transition of ettringite to monosulfate.

As shown in Figure 9.3, ettringite needle formation causes a significant change in microstructure when monosulfate is converted. The new microstructure is clearly more porous. When the sulfate ions are

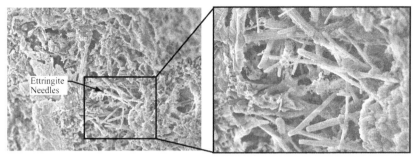

FIGURE 9.3. The action of sulfates increases pore sizes in concrete, creating opportunities to inject species that provide other benefits such as anti-biological agents. Adapted from Kunal Kupwade-Patil, 2010 with permission.

removed from the concrete, the degradation slows down. Once the existing ettringite needles have been saturated with absorbed water, the damage due to expansion can be expected to stop.

Because the microstructure now contains larger pores and microcracks that can be considered an extension of porosity, the opportunity exists for injecting other species into this altered pore structure—for example, loading the expanded pores with biocide particles such as copper oxide or a titanium dioxide. Since pore modification due to sulfate attack left the concrete in a weaker state, other nanoparticles of silica can be incorporated in order to recover strength.

This may seem outlandish, but it is conceivable that the intentional injection of sulfates could be accomplished in a calculated attempt to increase porosity temporarily. The altered pore structure could then be loaded with other beneficial species. Afterward, or even during the final stage of treatment, the sulfate ions still left in the pore fluid could be extracted by reversing the circuit polarity that was used to inject them.

9.3. POLYMERIC PHASE ASSEMBLY

The development of polymer-enhanced concrete has involved the use of dehydration, saturation with polymer precursors, and the application of gamma radiation to trigger polymerization. The results were impressive in terms of strength and durability; but lacked practicality. The following section describes an electrokinetic approach that can be used to avoid the use of dehydration and gamma radiation.

The key here is to find a polymer precursor somewhat soluble in water and carrying a net charge when dissolved. Methyl methacrylate

(MMA) is one such species. It is the precursor for poly methyl methacrylate (PMMA). A key feature of this monomer is that it can be induced to polymerize in the presence of a basic pH.

Figure 9.4 illustrates a treatment process involving MMA (Nayeem and Cardenas, 2011). The monomer is slightly soluble in water. It carries a negative charge that permits the use of electrokinetic transport into the pore structure. Driving this species directly to the reinforcement presents an interesting opportunity. The monomers at a very low dilution of say 1–2% will have relatively little opportunity to make contact with each other while in transit to the reinforcement. This reduces the chances of premature polymerization that could effectively limit a significant amount of monomer from loading.

Once a monomer arrives at the reinforcement, it must wait for another monomer to arrive before the elevated pH of the pore fluid can trigger a polymerization reaction. Similarly, this newly polymerized pair will wait for the next arrival before polymerization continues. In this way, the polymerization process can start at the reinforcement and continue growing fibers of polymer back through the pore.

In a preliminary PMMA synthesis trial conducted with the assistance of James Eastwood at the applied electrokinetics laboratory at Louisiana Tech University in September of 2008 the current density combined with the specimen geometry led to the curious extrusion of fibers from the top of the specimen (Figure 9.5). These submicron fibers grew in less than a week of treatment using an extremely high current density of 5 A/m².

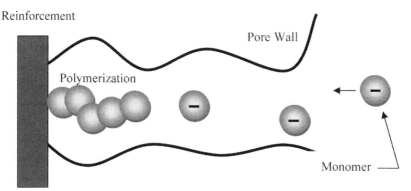

FIGURE 9.4. *Methyl methacrylate is slightly soluble in water and carries a negative net charge. When it is drawn to the reinforcement, the elevated pH of the concrete pore fluid causes it to polymerize.*

FIGURE 9.5. *Fibers of PMMA ranging from 10–100 microns extruded from the surface of a specimen of hardened cement paste during an electrokinetic treatment.*

Another trial conducted with a reduced current density of 1 A/m^2 and a treatment time of 12 days resulted in no extrusion of fibers (Nayeem and Cardenas, 2012). The tensile strength of the hardened cement paste increased 50% from 950 psi to 1500 psi. Figure 9.6 shows a fracture surface in which a fiber can be seen protruding from a pore.

The implications of synthesizing and growing polymers within the pores of concrete are interesting. Whisker formation of most materials tends to cause unusually high strength to develop, because the defect concentration in a whisker of material is often lower than that achieved in a bulk process. In the current case, the opportunities for defects to exist in the cement pores is quite low, due to the isolated and protected nature of the environment. In fact the electric field drawing methyl methacrylate toward the interior of the specimen is also drawing most ions out of solution and toward either pole of the circuit. This can clearly increases the purity of the processing environment. A similar analysis to that conducted in Section 7.4 was used to estimate the tensile strength of the PMMA phase formed in this case. The preliminary estimate indicated a tensile strength of 19 ksi for these whiskers, 80% higher than is typical for PMMA. While this increase in strength may seem surprising, it is certainly consistent with the differences in strength observed when whiskers of materials are compared with bulk strength (Stepanov, 1995; and Berezhkova, G., 1973). Material whiskers, being relatively low in defects, have been shown to exhibit tensile strengths that are a factor of 2–10 higher than the bulk material.

The other interesting implication of this concept has to do with corrosion protection. The first line of defense is typically a surface coating. Since no such coating is defect free, it is a matter of time before aggressive species such as chlorides get past a surface barrier that is typically less than 0.005 in thickness. A PMMA formation would require minimal to no surface preparation, and would provide a barrier coating that ranges 2 inches from the reinforcement to the concrete surface. As this barrier forms, it also displaces the water out of the pore system, starting at the rebar interface and working outward. The solvated PMMA ions can almost certainly penetrate some of the smallest (1–2 nm) pores of hardened cement paste. This level of pore saturation would have a dramatically suppressive impact on corrosion processes at the reinforcement.

A couple of interesting dilemmas confront the prospect of forming PMMA at the concrete reinforcement. One concern is the polarity required to drive negative ions to the reinforcement—a polarity that will cause the iron reinforcement to corrode. The second concern is the need for an elevated pH to trigger the polymerization process. Older concrete loses high pH due to the natural processes of carbonization and leaching.

A re-alkalizing pre-treatment of the steel reinforcement can remedy

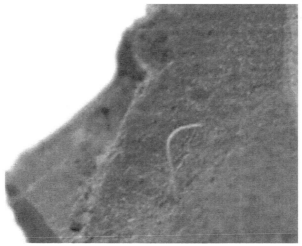

FIGURE 9.6. Single fiber of PMMA emanating from a pore on a fracture surface of a hardened cement paste specimen.

both concerns. Re-alkalizing concrete involves electrokinetic treat-ments using calcium, sodium, and potassium hydroxides, in which vari-ous oxides of these metals are electrodeposited on the surface of the re-inforcement. It was noted earlier that a combination of all 3 hydroxides can be used to form a sacrificial coating on iron reinforcement, while the polarity needed to drive MMA into the concrete removes it. Re-alkalization also resolves the pH concern, because the formation of the ceramic protective layer causes the pH in the vicinity of the reinforce-ment to rise substantially. This condition thus provides the pH needed for the precursors to polymerize as they reach the reinforcement.

9.4. NANOCOMPOSITE PHASE ASSEMBLY

This approach is based in part on the principal of electrostatic lay-er-by-layer assembly achieved by alternate adsorption of oppositely-charged nanoparticles and/or ions. In this case, the process is actually more electrodynamic in nature because weak electric fields are used to transport the charged species into the pores of the cementitious mate-rial. Various combinations of treatments were conducted over a period of 7 to 14 days. The following sections describe how these treatments were carried out and the interesting characteristics that they produced in concrete masonry blocks.

9.4.1. Combining Modes of Assembly

Layer-by-layer (LbL) self-assembly by alternate adsorption of op-positely-charged components (polyelectrolytes, nanoparticles and pro-teins) was developed in 1998 as a simple and versatile nanotechnology for thin coating surfaces (Lvov, Decher and Möhwald, 1993; Ichinose, Lvov and Kunitake, 2003; and Decher, 1997). These assemblies are usually in the vicinity of 10–500 nm. In related work, LbL assembly was conducted within 500 nm alumina pores for template synthesis of nanotubes (Li and Cui, 2006). No attempt had been made to apply this nanoassembly method to ordinary Portland cement until 2007.

In traditional LbL assembly, sequential adsorption of oppositely-charged components is carried out on a solid surface, such as a glass slide. The substrate is alternately immersed solutions or suspensions of polycations and negative nanoparticles. This mode of assembly is now combined with electrokinetic transport in order to create a reaction zone within the pores of concrete. In some cases, different species known to react on contact were driven toward each other to meet inside the pores and form new phases. The treatments were repeated to increase thicknesses of various phases in an attempt to increase strength. If thick films and possibly new phases can be built within a porous material, it is then possible to increase the tortuosity of the pore system, reduce porosity, and enhance strength. The following sections describe the methods used to attempt these outcomes.

9.4.2. Assembling Phases in Masonry Block

Emphasis was placed on examining the impact of varied combinations of reactive and charged ions and nanoparticles. Each treatment was executed using a controlled current density of 1 A/m^2. Various combinations of sodium silicate, calcium hydroxide, poly diallyl dimethyl ammonium chloride (PDDA) chains, acrylic nanoparticles and alumina-coated silica particles were transported into the concrete pores and toward other oppositely charge species with electrostatic assemblies and possibly some other reactions occurring.

The American Block Corporation of Bossier City, LA. provided lightweight cementitious masonry blocks (Cardenas *et al.* 2008). Masonry sealant was applied to the top and bottom of each block in order to provide a sound substrate application of window putty. The putty was used to create a seal between the bottom of each block and the floor of its treatment container. In this way, fluid from one side of the block was prevented from seeping under it and mixing with the other treatment fluid. The treatment circuit is illustrated in Figure 9.7. The window putty was provided by Old House Inc., Gardiner ME. Mixed metal oxide-coated titanium wire was used for connection to the positive pole of the circuit. The negative pole of the circuit was connected to zinc-plated steel wire cloth. The cloth mesh opening was 0.5 cm and had a 3 mm wire thickness.

The various chemical species used for treatments are listed in Table 9.1. Sodium silicate was used because it exhibited a broad range of ions and particle sizes, and is known for its pozzolanic reactivity. PDDA polymer is a positive polymer with chain structure. It is a well-characterized species, frequently used in LbL assembly work. The acrylic co-polymer carried a negative charge and particle sizes of 60 nm. The alumina-coated silica carries a positive in charge and a particle size of 24 nm.

A description of the treatment protocols used is found in Table 9.2.

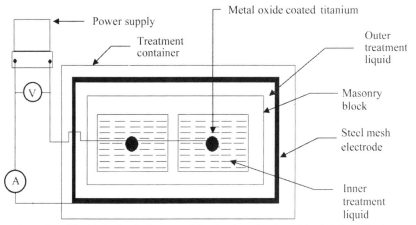

FIGURE 9.7. *Electrical circuit diagram of the treatment setup. Adapted from Cardenas H., Lvov, Y., Kurukunda, A., (2008) Electrokinetic Assembly of Polymeric and Pozzolanic Phases within Concrete. Reprinted by permission from the Society for the Advancement of Materials and Process Engineering (SAMPE).*

This table defines the shorthand notation used to refer to each protocol. Arrangement and connections of the power supply leads are illustrated in Figure 9.7. The location of a treatment fluid depended upon the sign of the charge carried by the treatment species driven into the masonry block wall. In all cases, the negative pole of the power supply was connected to the end of the steel mesh electrode surrounding the exterior of each block. Negatively-charge species were always loaded adjacent to this electrode. Similarly, the positive pole was connected to the mixed metal oxide-coated titanium electrode that was located in both the interior chambers of the masonry block. These chambers were always loaded with treatment species that carried a positive charge.

Following each treatment, maximum load-resistance tests were con-

TABLE 9.1. Treatment Agents.

Agent	Size	Charge	Weight Percent of Fluid (%)
Calcium hydroxide	Ionic	Positive	2
Sodium silicate	Variable size	Negative	44
PDDA polymer	Variable size	Positive	0.06
Li stabilized silica	50 nm	Positive	10
Acrylic co-polymer	60 nm	Negative	38
Alumina-coated silica	20 nm	Positive	30

TABLE 9.2. Specimen Treatment Protocols.

Protocol Short Hand Notation	Treatment Time (days)	Description
(SS+CH)·7	7	Treated with sodium silicate and calcium hydroxide for 7 days
(SS+CH)·7-IM	7	Same as above but by immersion only—no electric current
[(SS+CH) →7·(PDDA+SS) ·7] ·14	14	Sodium silicate and calcium hydroxide for 7 days followed by PDDA and sodium silicate for 7 more days
[(SS+CH) ·2↔(ACRYLIC+AL)·2]·14	14	Treated with two different sets of liquids: sodium silicate with calcium hydroxide and acrylic co-polymer with alumina- coated silica. Each pair was alternated every 2 days over a 14 period
[(SS+CH)·2↔(Li+CH)·2]·14	14	Treated with two different sets of liquids: sodium silicate with calcium hydroxide and lithium-stabilized silica with calcium hydroxide. Each pair was alternated each every 2 days over a 14 period
(ACRYLIC+AL)·7	7	Treated with acrylic co-polymer and alumina-coated silica for 7days
(ACRYLIC+CH)·7	7	Treated with acrylic co-polymer and calcium hydroxide for 7days
(AL+CH)·7	7	Treated with alumina-coated silica and calcium hydroxide for 7 days

Adapted from Cardenas H., Lvov, Y., Kurukunda, A., (2008) Electrokinetic Assembly of Polymeric and Pozzolanic Phases within Concrete. Reprinted by permission from the Society for the Advancement of Materials and Process Engineering (SAMPE).

ducted in accordance with ASTM C496 (2011). Care was taken to ensure that the centroid of the bearing surface was well aligned with the centroid of the testing machine's thrust plate. After load testing, porosity testing was also conducted in selected cases. Prior to the porosity test, each sample was stabilized in limewater. Each saturated sample was oven dried at 105°C. Porosity was calculated based on the loss in mass and the initial mass of the sample.

9.4.3. Compressive Strength Achievements in Masonry Block

Compressive load-resistance tests were conducted to assess treat-

TABLE 9.3. *Compressive Load Limits for Treated Masonry Block.*

Block Treatment	Failure Load* (lbs)	Increase in Strength (%)
Control	5600	—
(SS+CH)·7	11300	107
(SS+CH)·7-IM	10300	84
[(SS+CH) →7·(PDDA+SS) ·7] ·14	12000	111
[(SS+CH) ·2↔(ACRYLIC+AL)·2]·14	12400	120
[(SS+CH)·2↔(Li+CH)·2]·14	23300	316

*Values represent the average of three trials. Error range ± 5%.
Adapted from Cardenas H., Lvov, Y., Kurukunda, A., (2008) Electrokinetic Assembly of Polymeric and Pozzolanic Phases within Concrete. Reprinted by permission from the Society for the Advancement of Materials and Process Engineering (SAMPE).

ment protocol impacts. The load-resistance values obtained in each case are listed in Table 9.3. A 107% increase in strength was observed for blocks treated with sodium silicate and calcium hydroxide for seven days. When the same combination of treatment agents were used by simple immersion and no applied current, a strength increase of 84% was obtained. This may be due to treatment fluids simply flowing into macropores (over 1/8 inch) and being allowed to dry and solidify into the large voids. The sodium silicate, being of higher density, was probably the dominant filling agent. When dried, it forms a strong glass, but unfortunately can readily re-dissolve. Conceivably the electrokinetic version of this treatment would have provided sufficient calcium ions to cause a conversion to C–S–H which is not soluble.

Using 2 different pairs of treatment agents required a total of 14 days of treatment (Cardenas et al. 2008). In the first such case considered, one set of treatments was run for 7 days followed by another pair run for 7 additional days. Blocks treated with sodium silicate and calcium hydroxide for one week and PDDA polymer and sodium silicate for another week exhibited a 111% increase in strength. In the next case the same two pairs of treatment agents were run in a series of 2 day intervals. The treatments were repeatedly switched every two days instead of just once after 7 days. The blocks treated by this 2-day alternating approach using the same pairs of agents showed an average increase of 120% in strength as compared to the controls. One of the treatment protocols was found to exhibit a more exceptional impact on strength. Blocks treated with the lithium-stabilized silica protocol exhibited a strength increase of 316%. It is conceivable that this treatment may have produced a high-strength lithium silicate glass.

9.4.4. Porosity Reductions in Masonry Block

The porous volume contents exhibited by masonry block units are shown in Table 9.4. Every treated specimen incurred a significant porosity reduction. The case involving sodium silicate/calcium hydroxide alternated with acrylic co-polymer showed the highest porosity reduction—52%. This reduction may be due in part to a nanoscale polymer film formation enhanced by the electrodeposition of the 60 nm acrylic co-polymer particles. Electrokinetically-assisted film formation would be expected to close the interparticle spaces that otherwise exist in a specimen that is simply immersion-loaded in a nanoparticle bath.

Some of the porosity reduction observed may be due to the formation of C–S–H and C–A–H within pores. This assembly mode would stem from reactions between alumina or silica nanoparticles and the calcium hydroxide abundantly available, either from resident deposits or from the treatments. In addition, it is also possible that coagulated nanoparticles assembled at bottlenecks, where they collected and formed locally dense phases.

This work sought to examine another possibility. It was intended that the coagulation of oppositely-charged particles occurred where they were driven into contact by the electric field. They arrived from either side of the masonry block wall, stuck together through electrostatic attraction, and fell out of suspension within the pores. These oppositely-charge species could thus assemble in a manner similar to layer-by-layer assembly. Wherever these electrostatically coagulated species dropped would be the start of a new potential obstruction, upon which new arrivals would collect to form densified assemblies within the pores of the old material.

Another assembly mode touched on earlier is that of polymer particle

TABLE 9.4. Treatment Impact on Porosity of Masonry Block Specimens.

Blocks	Control Porous	Volume Fraction	Porosity Reduction (%)
(SS+CH)·7	0.109	0.187	43
[(SS+CH) →7·(PDDA+SS) ·7] ·14	0.117	0.175	33
[(SS+CH) ·2↔(ACRYLIC+AL)·2]·14	0.091	0.190	52

Adapted from Cardenas H., Lvov, Y., Kurukunda, A., (2008) Electrokinetic Assembly of Polymeric and Pozzolanic Phases within Concrete. Reprinted by permission from the Society for the Advancement of Materials and Process Engineering (SAMPE).

self-assembly into films. This takes two steps. The first requires the coagulated particle assembly to dry. As water evaporates, the polymer chains of each particle are free to unravel and interact with neighboring particles, gradually forming a film that reduces pore connectivity and permeability. It is conceivable that such films could also promote increases in tensile as well as compressive strength.

Following the compression tests, selected specimens were examined via SEM. Figure 9.8 is an image of an untreated masonry block. Figure 9.9 shows a similar image of a masonry block specimen that received a 7-day treatment of sodium silicate and calcium hydroxide. The treated sample appears to show less porosity than that observed in the control specimen in Figure 9.8. This is not a very surprising outcome, because the treated specimens had a 43% porosity reduction compared with the controls—which translated to a 107% increase in strength.

Figure 9.10 contains an atomic force microscope image of a sample removed from a block specimen that was treated with 60 nm copolymer particles. The image is an aspect representation of the specimen surface. The vertical axis displays nanoscale resolution. The horizontal axes are scaled at the micron level of resolution. There are a large number of peak elevations in the vicinity of 25–30 nm, indicating a region of de-

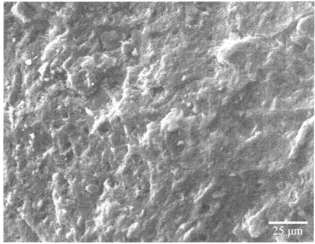

FIGURE 9.8. SEM image of an untreated block. Reproduced from Cardenas H., Lvov, Y., Kurukunda, A., (2008) Electrokinetic Assembly of Polymeric and Pozzolanic Phases within Concrete. Reprinted by permission from the Society for the Advancement of Materials and Process Engineering (SAMPE).

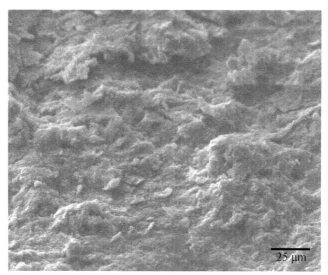

FIGURE 9.9. SEM image of the block treated with sodium silicate and calcium hydroxide. Reproduced from Cardenas H., Lvov, Y., Kurukunda, A., (2008) Electrokinetic Assembly of Polymeric and Pozzolanic Phases within Concrete. Reprinted by permission from the Society for the Advancement of Materials and Process Engineering (SAMPE).

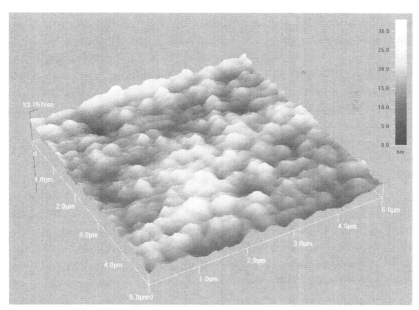

FIGURE 9.10. Self-assembly in acrylic copolymer. Reproduced from Cardenas H., Lvov, Y., Kurukunda, A., (2008) Electrokinetic Assembly of Polymeric and Pozzolanic Phases within Concrete. Reprinted by permission from the Society for the Advancement of Materials and Process Engineering (SAMPE).

165

posited nanoparticles that show their half-heights. Three of the peaks are at the 30 nm elevation, while most of the others are at 25. There are apparent valleys in the image as well. At these locations of 0–5 nm elevation, no particle could fit in the deposit. Another possibility is that some of the nanoparticles have engaged in film formation, which would reduce the half-heights of the peaks. Since film formation would engage more than one particle, it makes sense that some multi-particle regions exhibited reduced elevation. This could be the case in parts of the image where an elevation of the 10 nm height appears to involve more than one nanoparticle site. This pattern appears to indicate some amount of peak reduction, possibly associated with film assembly. While porosity reduction may not be directly impacted by this behavior, the impact on permeability would probably be significant.

9.4.5. Summary of Nanocomposite Assembly

This section examined the combination of two nanotechnology methods: polyelectrolyte layer-by-layer nanoassembly and electrokinetic nanoparticle delivery into cementitious materials. Each treatment succeeded in reducing the porosity and increasing strength. Polymer film assembly was observed using particles transported into pores of concrete masonry block. A lithium-stabilized silica particle was electrokinetically injected, in combination with sodium silicate and calcium hydroxide. These treatments yielded remarkable strength enhancement to the masonry blocks, with increases topping 300%.

9.5. LITHIUM-COATED SILICA TREATMENT

A particularly interesting microstructural change was obtained using 50 nm lithium-coated silica particles (Cardenas *et al.* 2009). Sections of concrete 12 inches deep were treated in an attempt to load the pores with a source of lithium. Extremely high current densities were used in order to ensure that each treatment lasted less than 12 days.

Figure 9.11 shows the impact of the high current treatments using 50 nm silica-coated 2 nm lithium oxide particles. The images on the right side of the figure show typical morphology of untreated concrete. The left side shows the impact of high current combined with this nanoparticle. The microstructure was significantly altered.

Table 9.5 shows how each section of 12-inch treatment influenced

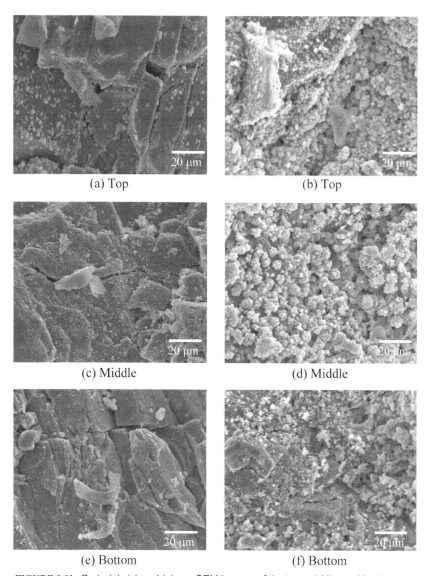

(a) Top

(b) Top

(c) Middle

(d) Middle

(e) Bottom

(f) Bottom

FIGURE 9.11. Parts (a), (c) and (e) are SEM images of the top, middle, and bottom sections of a specimen treated at high current density (10 A/m²); (b), (d), and (f) are the SEM images of control specimens that were not treated (Cardenas, 2009).

the strength of the concrete. In typical chloride extraction processes, the current density used is at or below 1 A/m² in order to avoid damage leading to a reduction of strength in the concrete. Despite the extraordinarily high current densities utilized, it was clear that all specimens in-

TABLE 9.5. Strength Increases Using Ultra-High Treatment Current (Cardenas, 2009).

Location of Sample	Low Current [3 A/m^2]	Medium Current [6 A/m^2]	High Current [10 A/m^2]
Top	56%	50%	32%
Middle	12%	22%	33%
Bottom	23%	16%	36%
Average	30%	29%	34%
Error range	±9%	±8%	±6%

*Each value is an average of 3 section results.

creased in strength, and that the nanoparticles were able to compensate for any damage caused by the high current density.

This treatment provided new insight into deeply-penetrating lithium delivery, the use of extremely high processing currents, and the significant alterations that electrochemical treatment can achieve in the microstructure and strength of concrete. These radical alterations may possibly be associated with the in situ formation of lithium silicate glass. All in all, a compelling example of an electromutagenic process applied to concrete.

9.6. OPPORTUNITIES

The methods and discoveries presented so far only scratch the surface of the opportunity we have to improve concrete with the application of nanomaterials. As these nanotechnology methods are developed and applied, it is intriguing to consider how many concrete facilities will be spared demolition as old structures are repaired from within, micro-cracks are healed, pore structures enhanced, strength increased, and durability improved.

In thermal shock, water adjacent to the surface flashes into steam, and the rapid discharge of this steam creates cracks. Alumina nanoparticles can form a high-alumina cement inside pores of existing concrete to enhance thermal resistance. Later, the conversion of these phases to more stable and dense versions, permits sufficient porosity for the rapid escape of water vapor.

Another application is in the area of abrasion resistance, which correlates well with strength. Because the strength of concrete grows as porosity is reduced, a nanoparticle treatment that reduces porosity increases strength. It is clear that the abrasion resistance of the material would also benefit.

Alkali silica reaction (ASR) occurs when reactive forms of aggregate start forming a hydroscopic gel. Ensuing expansion causes serious dimensional instability of the structure, as well as cracking. The keys to mitigating this include keeping additional water out and providing a chemical species that stops the ASR gel from expanding. Nanoparticle treatment can reduce porosity enough to effectively deny water from accessing the concrete. An additional treatment, delivering lithium, calcium or other ions that are known to cause the ASR gels to become dimensionally stable, would also help.

The ability to deliver nanomaterials to concrete structures where they stand is a key weapon in the fight to maintain our national infrastructure. It is not only an extremely promising way to upgrade structures that are being considered for demolition, but also a means to practically reinvent existing structures by providing them with new interior nanostructure and microstructure. A structure nearing its 50–75 year design life can be upgraded so that its life extends another 50–75 years. We can equip the interior of our structures with nanocomposite precursors that effectively turn old structures into new ones. We can load nanomaterials in various combinations and stages that allow old structures to exhibit higher strengths than ever imagined. These new nanostructures can be designed to respond to changes in the environment. For example, should micro-cracking become a problem due to unforeseen changes in the service environment, nanomodified structures will have the ability to release epoxy constituents from neighboring stages of electrokinetic deposits, which will combine and provide healing of the micro-cracks. Other layered electrokinetic deposits will release stored reserves of re-alkalizing agents to re-elevate the pH of old structures suffering environmental damage due to carbonization or other distresses. The application of nanomaterials to concrete structures will enable us to virtually rebuild them from within, upgrading our infrastructure exactly where it stands, without tearing it down and starting over.

9.7. REFERENCES

ASTM C496, "Standard Test Method for Splitting Tensile Strength for Cylindrical Concrete Specimens", ASTM International, West Conshohocken, PA, 2011.

Berezhkova, G.V., (1973) Monokristal'nye Volokna I Armirovanye imi Materialy", Moscow, (Translated from English).

Bonen, D., "Calcium Hydroxide Deposition in the Near Interfacial Zone in Plain Concrete", *Journal of the American Ceramic Society,* Vol. 77, No. 1, pp. 193–196, January 1994.

Cardenas, H. E., Syed, F., Eklund, S. E., *Electromutagenic Process Permits High Current Density for Lithium Transport in Concrete,* J. Construction and Building Materials, submitted July 2009.

Cardenas, H., Lvov, Y, Kurukunda, A., "Electrokinetic Assembly of Polymeric and Pozzolanic Phases within Concrete", *Proceedings, SAMPE 2008, Long Beach, CA,* 18–22 May, 2008.

Cardenas, H., Paturi, P., Dubasi, P., "Electrokinetic Treatment for Freezing and Thawing Damage Mitigation within Limestone", *Proceedings of Sustainable Construction Materials and Technologies, Coventry England, Taylor & Francis Group,* UK 12 June 2007.

Decher, G. (1997) "Fuzzy Nanoassemblies: Toward Layered Polymeric Multicomposites," *Science,* Vol. 227, 1232–1237.

Fan, Y. et al. (1999) "Activation of Fly Ash and its Effects on Cement Properties," *Cement and Concrete Research,* Vol. 29, No. 4, April, pp. 467–472.

Gollop, R.S., Taylor, H.F.W., "Microstructural and Microanalytical Studies of Sulfate Attack I. Ordinary Portland Cement Paste", *Cement and Concrete Research,* Vol. 22, No. 6, pp. 1027–1038, November 1992.

Ichinose, I., Lvov, Y. and Kunitake, T. (2003) *Multilayer Thin Films Chap 8: Sequential Assembly of Nanocomposite Materials,* eds. G. Decher, J. Schlenoff, J-M. Lehn. Wiley-VCH, Weinheim, p.155–176.

Kakali, G., Tsivilis, S., Aggeli, E., Bati, M., "Hydration Products of C3A, C3S and Portland Cement in the presence of $CaCO_3$", *Cement and Concrete Research,* Vol. 30, No. 7, pp. 1073-1077, 7 July, 2000.

Li, J. and Cui, Y. (2006), "Template Synthesized Nanotubes Through Layer-by-Layer Assembly Under Charge Interaction," *J. Nanoscience and Nanotechnology,* Vol. 6, 1552–1556.

Lvov, Y., Decher, G. and Möhwald, H. (1993) "Assembly, Structural Characterization and Thermal Behavior of Layer-by-Layer Deposited Ultrathin Films of Polyvinylsulfonate and Polyallylamine," *Langmuir,* Vol. 9, 481–486.

Kupwade-Patil, K., "Mitigation of Chloride and Sulfate Based Corrosion in Reinforced Concrete via Electrokinetic Nanoparticle Treatment," Ph.D. Thesis, Louisiana Tech University, Ruston, LA, August 2010.

Odler, I., Colan-Subauste, J., "Investigations on Cement Expansion Associated with Ettringite Formation", Cement and Concrete Research, Vol. 29, No. 5, pp. 731–735, May 1999.

Mehta P., and Wang, S., "Expansion of Ettringite by Water Absorption", *Cement and Concrete Research,* Vol. 12, No. 1, pp. 121–122, January 1982.

Mindess, S., Young, J. F. and Darwin, D. (2003) *Concrete* 2nd Ed., Prentice-Hall Inc., Pearson Education, Inc., Upper Saddle River, New Jersey.

Nayeem, N., Cardenas, H., (2012) Electrosynthesis of Polymethyl Methacrylate within Hardened Cement Paste," *Proceedings of the Green6 Conference, Anglia Ruskin University,* submitted Nov. 2011

Rasheeduzzafar, O. et al. (1994) "Magnesium-Sodium Sulfate Attack in Plain and Blended Cements," *J. Mater. Civ. Eng.,* 6(2):201–222.

Taylor, H. F. W. (1990) *Cement Chemistry,* 2nd Edition, Thomas Telford Pub., London.

Sanchez de Rojas, M. and Frias, J. (1999) "Influence of the Microsilica State on Pozzolanic Reaction Rate," *Cement and Concrete Research,* June, Vol. 29, No. 6, pp. 945–949.

Solberg, C., Hansen, S., "Dissolution of $CaSO_4 \cdot 1/2H_2O$ and Precipitation of $CaSO_4 \cdot 2H_2O$, A Kinetic Study by Synchrotron X-ray Powder Diffraction", *Cement and Concrete Research,* Vol. 31, No. 4, pp. 641–646, April 2001.

Taylor, H. F. W. (1997) *Cement Chemistry,* 2nd Edition, Thomas Telford Pub., London.

Zivica, V. "Sulfate Resistance of the Cement Materials Based on the Modified Silica Fume", *Construction and Building Materials,* Vol. 14, pp. 17–23, Elesevier Science Ltd, 2000.

Index

171

About the Author

HENRY E. CARDENAS, PH.D. is the Jack T. Painter endowed Professor of Civil Engineering at Louisiana Tech University, Ruston. He is also Associate Professor in the Departments of Mechanical as well as Nanosystems Engineering. Connecting the dots across a range of disciplines has brought Dr. Cardenas international recognition for his award winning research in which nanomaterials are being applied to a variety of practical problems. He is the director of the Applied Electrokinetics Laboratory in the College of Engineering and Science. With over 25 years of experience in Defense, industrial and academic research, he has over 40 publications in the area of materials performance and durability.